Author's Acknowl

Readers of this study are not likely to surmise the extent of my indebtedness to numerous individuals in the communications industry and in government for their willingness not only to answer numerous questions but also to comment freely on tentative conclusions that I reached in the course of the research. Since these individuals shared their views on a confidential basis, they must remain anonymous. If this study should have beneficial consequences for the industry, the government, and the public, a substantial share of the credit must go to these individuals.

A number of my faculty colleagues read an early draft of the manuscript. Some of them raised specific questions and made constructive suggestions, thus contributing significantly to the final version. David L. Birch and Lewis M. Schneider were particularly helpful in this respect, as was Mark S. Massel, formerly with the Brookings Institution and author of *Competition and Monopoly* (1962). Others expressed serious doubts regarding the merits of the approach taken in this study. In an effort to take cognizance of their observations, I had to make explicit assumptions and conclusions which had been merely implicit in the earlier draft. Credit for these revisions, therefore, must go to the colleagues who had the courage to voice their doubts.

I am grateful to my secretary, Mrs. Dorothy Morgan, for her competence and cheerfulness in retyping the manuscript as it went through several drafts, and to Ruth Norton for skillfully converting the final manuscript into a neat volume.

I appreciate the continuing encouragement of Professor Bertrand Fox, who was Director of the Division of Research when this study began, and to Professor Lawrence E. Fouraker, the present Director. I wish to thank Dean George P.

Baker, Associate Dean George F. F. Lombard, and former Area Chairman Paul W. Cherington for letting me proceed in my own way in undertaking this study.

Finally, I wish to thank my wife, Narnie, for acting as a long-suffering but sympathetic sparring partner in discussing and questioning many of the ideas expressed in this volume.

The presentation and conclusions contained in this study have been revised and re-revised under the impact of views expressed to me in numerous comments and discussions. Without them, this study would not have been possible. However, the sole responsibility for the final product in the pages that follow is mine.

KURT BORCHARDT

Soldiers Field
Boston, Massachusetts
November 1969

Structure and Performance of the U.S. Communications Industry

"All intelligent political criticism is comparative. It deals not with all-or-none situations, but with practical alternatives."

John Dewey
The Public and Its Problems (1927)

Structure and Performance of the U.S. Communications Industry

Government Regulation and Company Planning

KURT BORCHARDT
Lecturer on Business Administration
Harvard University

DIVISION OF RESEARCH
GRADUATE SCHOOL OF BUSINESS ADMINISTRATION
HARVARD UNIVERSITY

Boston • 1970

Library of Congress Catalog Card No. 73–104322
Standard Book Number 87584–082–5

Faculty research at the Harvard Business School is undertaken with the expectation of publication. In such publications the author responsible for the research project is also responsible for statements of fact, opinions, and conclusions expressed. Neither the Harvard Business School, its Faculty as a whole, nor the President and Fellows of Harvard College reach conclusions or make recommendations as a result of Faculty research.

Printed in the United States of America

ABSTRACT

Structure and Performance of the U.S. Communications Industry: Government Regulation and Company Planning

The future structure of the electronic communications industry has important implications for government, business, and consumers — indeed not only for the entire economy but for social and political patterns as well. This study by Dr. Borchardt is directed at evaluating and managing the structural problems of this industry that result from the interaction of public regulation and rapidly changing technology. It suggests modifications that need to be made in the regulatory procedures to ensure an efficient industry in an environment of new technologies and growing demands for increasingly diverse services.

Dr. Kurt Borchardt is a Lecturer on Business Administration at the Harvard Business School. His extensive background in handling communications legislation for a period of twenty years as legal counsel for the Interstate and Foreign Commerce Committee of the U.S. House of Representatives provided him with a useful background for undertaking this research.

This study seeks to identify "structural" problems in the U.S. communications industry and to suggest approaches for handling them more effectively and expeditiously. Structural problems present themselves when the twin questions to be decided upon are: Who does what? Who gets what? The study traces how these questions have been answered over a period of 125 years and how the present configuration of the electronic communications industry has resulted from unilateral actions of individual companies, intercompany agreements, and government regulation. From here on governments rather than company managers are likely to have the greatest impact with regard to what any company or group of companies may do and what rewards they will reap unless company managers are prepared to exercise greater initiative in generating acceptable answers to these questions.

New technologies and new services made possible by them not only tend to blur longstanding divisions of functions within the communications industry; they also render uncertain the borders between that industry and the computer industry. The resulting structural conflicts between individual companies and industry segments in these two industries are interconnected like the nodes in the web of a spider. Time and energy-consuming major battles are fought over such issues as (1) the extent to which customer-owned equipment may be attached to and private communications systems may be interconnected with the switched telephone network; (2) the extent to which independent manufacturers of communications equipment may secure access to the huge equipment market which at present is supplied largely by integrated telephone companies; and (3) the extent to which bulk users of communication services may provide their own services or secure them from specialized carriers which seek to enter into competition with the regular carriers.

Failure of present government machinery and procedures to resolve conflicts of this kind with reasonable speed and effectiveness has resulted in growing frustrations for both business managers and government officials. Efforts to develop definite public policies to guide regulatory agencies in allocating tasks and opportunities among the feuding parties necessarily have come to naught. Under these circumstances a combination of broad regulatory goals and informal procedures designed to achieve these goals are urged by the author. The goals which stand in a relationship of tension to one another are: to preserve incentives and capacities of individual companies to plan their future roles in the communications system, and at the same time to improve, or at least not affect adversely, the future performance of the communications system as a whole. As the number of companies increases and the companies become more differentiated, company managers must learn to serve two masters: the communications system and their own companies. Government regulators, on the other hand, must focus on the long-range impact of particular allocations of tasks and opportunities on the incentives and capacities of companies to be progressive and dynamic organizations.

The magnitude and interrelatedness of the structural problems and the inability of traditional government machinery and procedures to cope with these problems have led to sporadic efforts to resolve the conflicts on the macro-political level where the President and his immediate staff assistants become involved. The handling of structural problems in this country is compared with ways of handling them in Canada. The creation of a Department of Communications and use of more flexible and informal procedures have made possible trade-offs in dealing with interrelated problems conserving time and energy of company managers and government officials. Following a similar approach in the United States requires creation of a permanent capability on the macro-political level which can exercise effective leadership and coordination. Such an approach is now under study but higher priorities of other domestic and foreign problems may lead to a postponement in creating such a capability either in one of the present departments or in the Executive Office of the President.

Permanent escalation of structural problems to the macro-political level raises the basic question how the decision-making function can be properly allocated between the private and public sectors. The author urges company managers to create new intercompany and industry-government mechanisms modeled after the Atomic Industrial Forum which will follow a systems approach in studying and evaluating alternative allocations of tasks within the communications system. While such mechanisms will not eliminate intercompany or industry-government conflicts, they will provide rational inputs which are likely to narrow the scope of controversies and achieve a better balance between industry and government in deciding structural problems.

The author compares the solutions advanced to solve structural problems in the communications industry with solutions of such problems in the transportation industry. He suggests that "competition" between, rather than adequate coordination of transportation functions performed by, different companies engaged in different modes of transportation has necessitated reliance on the Federal Government to achieve such coordination through the recently created Department of Transportation. The role of the Department of Justice under the antitrust laws in shaping the present industry structure is examined in some detail, and it is shown how at various times the Department and other government agencies have been in fundamental disagreement over whether "competition" or "coordination" is called for in the public interest. A number of situations are discussed in which the strong desire to avoid monopolies and yet not to depend wholly on public utility-type regulation has led to efforts of building into the industry structure a maximum number of "competitive" features. However, there is no actual market place where the competing companies may challenge each other; instead "competition for regulation" takes place before government agencies.

The author stresses that our present knowledge of relations between industry structure and industry performance is extremely limited, especially under rapidly changing technologies. In shaping the future structure of the communications industry we must have regard for the interactions between *knowledge* supplied by different disciplines, *organizations* which apply this knowledge, and *procedures* employed in applying it. Interactions between these three elements and not each of these three elements separately make the decision-making processes what they are.

In closing the author points out that in the future we shall need a greatly increased supply of personnel capable of acting not only as interpreters and integrators of different disciplines but also as interpreters and integrators of the privately organized and publicly organized sectors of our political economy. He appeals to academic institutions to help train such personnel and thus contribute to achieving a proper allocation of decision making between these two sectors.

(Published by Division of Research, Harvard Business School, Soldiers Field, Boston, Mass. 02163. xi + 180 pp. $6.00 1970)

Foreword

The future structure of the electronic communications industry has important implications for government, business, and consumers — indeed not only for the entire economy but for social and political patterns as well. Dr. Borchardt's study is directed at evaluating and managing those structural problems in the U.S. communications industry that result from the interaction of public regulation and rapidly changing technology. Dr. Borchardt traces the evolution of the present communications system, emphasizing the principles that have been applied in apportioning that system among various producers. He projects the consequences of these apportioning principles in an environment of changing technologies and growing demands for increasingly diverse services. The question of what modifications need to be made in the regulatory procedures is then considered.

Dr. Borchardt's extensive experience in handling communications legislation for a period of twenty years as legal counsel for the Interstate and Foreign Commerce Committee of the U.S. House of Representatives provided him with a useful background for undertaking this research. Further, he carried on discussions with numerous public and private officials during the course of the study.

This research project was made possible by a grant from the Alfred P. Sloan Foundation and the income from the B. F. Goodrich Company Endowment for Research in Memory of David M. Goodrich. We are grateful to the donors for these resources.

LAWRENCE E. FOURAKER
Director, Division of Research

Soldiers Field
Boston, Massachusetts
November 1969

Table of Contents

PART II. PROCESSES WHICH DETERMINE THE STRUCTURE OF THE COMMUNICATIONS INDUSTRY

Introduction

CHAPTER I

Introduction

Purpose of the Study

The present configuration of the electronic communications industry in the United States has evolved over a period of 125 years. It is the result of unilateral actions by individual companies, intercompany agreements, and government regulation.

The present configuration is being challenged by various new technological developments. Coaxial cables and communication satellites, for example, tend to blur some of the traditional divisions of functions between different companies within the communications industry. The use of digital computers tends to render uncertain the present borders separating the communications industry from the data processing industry.

As new technologies make possible the development of new communication and information services, new challenges to the present structure of the communications industry can be expected. What will be the politico-economic framework within which intercompany and interindustry conflicts over the distribution of new business opportunities within and beyond the borders of the communications industry will be decided? What should be the goals of company planning and government regulation with regard to such distribution, and how should individual companies, industry associations, and government agencies seek to achieve workable regulation in this respect? These are the questions with which the present study is concerned.

The general premise on which this study is based has been well stated by Emmette S. Redford:

> The most important issue of the future is not whether government will surround the exercise of private power with restrictions and regulations, but whether the decision-making function will be properly allocated between the privately organized and the publicly organized sectors of the political economy.[1]

With this as its premise, this study seeks to explore (1) how the decision-making function with respect to the structure of the American communications industry has been exercised in the past, (2) how the publicly organized sector of the political economy is likely to exercise that function in the future, and (3) what the privately organized sector might do to promote the proper allocation between the two sectors.

In studying the proper allocation of the decision-making function between the private and public sectors, we shall focus on (1) *organizations,* both public and private, which participate in the decision making, (2) *reasoned arguments* based on explicit or implicit values supplied by various disciplines in support of alternative ways of structuring the industry, and (3) decision-making *procedures* employed by such organizations in the course of which reasoned arguments are considered.

A word of caution should be addressed at the outset to readers who expect elaborations, based on cost-benefit analyses, of optimal solutions of structural conflicts in the communications industry. In the first place, the study will seek to show there does not exist at present — and it is hoped that there will not be in the future — a single governmental decision-making agency empowered to direct the structuring of the communications industry in accordance with a particular model considered optimal by that agency. Second, there ap-

[1] Reference footnotes will be found in this book beginning on page 173.

pears to be a growing realization, at least on the part of some skilled system analysts, that cost-benefits analysis techniques have not been perfected sufficiently to produce optimum results when applied to complex politico-economic problems.

Two recent expressions of opinion are representative of this point of view. The Stanford Research Institute, under contract with the Federal Communications Commission, prepared a report entitled "Policy Issues Presented by the Interdependence of Computer and Communications Services." [2] The Institute sought to consider the principal regulatory alternatives and to assess some of the consequences and the resulting costs and benefits associated with the selection of each alternative. It concluded that the most serious weakness of the study was the Institute's inability to quantify the costs and benefits associated with taking any given action, and it therefore urged the FCC to continue on a step-by-step basis the moderate rate of change that it had set in motion in reaching decisions on particular aspects of the overall problem.

A second view was expressed by the president of the Lockheed Aircraft Corporation in an address entitled "Applying the State of the Art to the Problems of the State." [3] He stated that the aerospace industry had been successful in the past in applying systems analysis in fields where the problems could be well structured, where man — the complex social organism — did not constitute a major component in the system, but that a great deal of research and development would be required before the industry could employ the systems approach to such problems as decaying cities and the transportation crisis with any degree of certainty that it would do more good than harm.[4]

The purpose of the present study, therefore, is limited to providing an organizational, procedural, and conceptual framework within which interactional decisions can be reached and cost-benefit analysis techniques can be applied to the extent that they can contribute to reaching solutions in politico-economic conflicts between public and private groups in the communications industry.

Since the proper allocation of decision-making responsibilities between the private and public sectors presents difficult problems in other large politico-economic systems such as transportation, energy, finance, food, housing, and health, it is hoped that the framework developed in this study will be helpful also in dealing with those systems.

Scope of the Study

This study is concerned with the question of what constitutes workable regulation of the *structure* of the electronic communications industry. It is concerned with regulations designed to affect the content of radio and television programs only insofar as such regulations affect the structure of the broadcasting industry, one of the segments of the communications industry.

The description of the structure of an industry tells us what functions various members of that industry perform. In the case of a regulated industry like the communications industry, the performance of most functions is subject to governmental authorization. Companies which seek to supply communications hardware, and companies which seek to provide communication services either to others or for their own use in connection with noncommunication activities, enter into contests with each other over the respective functions which they should be authorized to perform. Structural regulation — and the term regulation here is used broadly — then is concerned with determining in the case of such contests which companies will be authorized to perform what functions.

Such contests are carried on not only before regulatory agencies such as the Federal Communications Commission. Congress also may be called upon to determine which company shall perform what functions. By enacting the Communications Satellite Act, Congress did so, for example, in the case of the Communications Satellite Corporation. Furthermore, the courts may make such determinations. A Federal court, for example, entered a decree in an antitrust case as a result of which the American Telephone and Telegraph Company (AT&T) was limited primarily to performing functions closely related to furnishing communication common carrier services. Finally, structural regulation in the broadest sense occurs, for example, when the government uses its procurement powers to promote the development or application of new technologies, or when it makes a conscious choice between becoming a large-scale user of services supplied by private companies and providing such services by means of government-owned facilities.

The term "structural regulation" is intended to distinguish regulations dealing with the division of functions among members of an industry from regulations concerned with the operations of those members after functional divisions have been made. However, the latter type of regulation may also influence in the longer run which companies will perform what functions, and thus determine the structure of an industry. For example, regulations concerning bulk rates charged by carriers for services and facilities are likely to determine whether customers will become self-suppliers of such services and facilities, and whether specialized carriers will seek entry into an industry. Therefore, the distinction is only relative and is essentially one which distinguishes regulations which are relatively infrequent and likely to have a more direct and permanent impact on the division of functions, and therefore on the structure of an industry, from regulations which occur with greater frequency and where their impact on industry structure is less direct and possibly less permanent.

Differences Between Several Categories of Electronic Communications

In order to comprehend the present configuration of the communications industry and the structural problems which confront that industry, it is essential that the differences between several categories of electronic communications be properly understood.

The first distinction which should be made — that between "intercommunications" and "mass communications" — is based on differences in the characteristics of the services provided and the facilities required by those two classes of communications. Intercommunication services enable individuals (or machines appropriately programmed by individuals) to communicate with each other, and the individuals furnish and control the content of the messages which are transmitted. Examples of intercommunication services are those provided by public message telephone, teletype, dataphone, or picturephone. The facilities required to furnish such services have the common characteristics of being bi-directional and of making possible individual selections of connections by means of appropriate switching devices.

In the case of mass communication services such as broadcasting or cable casting,* information (programs and advertising) is disseminated from a central point to numerous individuals. The disseminating companies or organizations control the message content but need not necessarily own the facilities required to furnish such services. The facilities have the common characteristics of being uni-directional, providing program selectivity (rather than individual selectivity),

* Cable casting is the dissemination by cable, for viewing on television receivers in the home, of programs which *originate* within the facilities of a cable (CATV) system. Cable casting is a service which may be performed by CATV systems in addition to the traditional dissemination by cable of programs originally broadcast by television stations and received over the air for such dissemination by special antennae operated by the CATV system.

and not requiring connections by means of switching devices among listeners, viewers, or subscribers to such services.

Within the class of intercommunications we shall have to distinguish between *public message* services, to which we have already alluded, and facilities used to furnish such services, and *private line* services and facilities used for such services. The latter are individually tailored intercommunication services and facilities. Sometimes the same private line facilities can be used for a variety of services such as telephone, telegraph, teletype, and data services.

The Evolution of the Present Structure of the Electronic Communications Industry

Timetable of Technological Developments

The present structure of communication companies within the communications system has developed largely as a result of (1) advances in communications technologies which have resulted in an increasing diversity of communication services being rendered by means of increasingly diverse modes of transmission; and (2) the distribution among various communication companies of business opportunities made possible by such technologies.

The most significant approximate dates on which new communication services utilizing new communications technologies were instituted are listed on page 10.

Intercommunication Companies

The present configuration of the communications industry reflects to a considerable extent the distinction between intercommunications and mass communications. Thus, we may generally distinguish between intercommunication and mass communication companies. When we focus on this distinction, however, we should not overlook the fact that mass communication firms, and particularly radio and television

Date	*Intercommunications*	*Mass communications*
1844	telegraphy by overland wires	
1866	telegraphy by undersea telegraphy cables	
1878	local telephone communications by means of wires and switchboards	
1880	long-distance telephone communications by means of overland wires	
1898	wireless ship-to-shore telegraphy	
1901	wireless transatlantic telegraphy using low radio frequencies	
1920		AM radio broadcasting
1927	transoceanic radio telephony using high radio frequencies	
1939	coaxial cable for long-distance overland transmission of various services	
1940		FM radio broadcasting
1941		television broadcasting
1944	point-to-point microwave for long-distance overland transmission of various services	
1950		coaxial cable for disseminating television programs (CATV)
1956	transoceanic telephony by undersea telephone cable	
1962	satellites for long-distance transmission of various services	

networks, as well as educational broadcasting stations, are large-scale users of intercommunication facilities, especially of long-distance intercommunication facilities. The three commercial television networks, for example, in 1967 paid the telephone carriers $45 million for interconnecting the networks and network-owned and affiliated stations. If increased new rates proposed by AT&T for television program transmission should become effective, the amount paid for television transmission will total $69.6 million by 1971. As heavy users of long-distance intercommunication services, mass communication firms and educational broadcasters therefore have become special claimants for opportunities to exploit possible savings from the use of domestic satellites.

The relative sizes of intercommunication companies vary greatly. While there are some 2,100 independent telephone companies in the United States, these companies operate only 17.2 million telephones as compared with the 87 million phones operated by the Bell System of the American Telephone and Telegraph Company. The second largest telephone system is the General System operated by General Telephone and Electronics Company with 8.7 million telephones, followed by United Utilities, Inc., Continental Telephone Corporation, and Central Telephone and Utilities Corporation, with 1.7 million, 1 million, and 0.76 million telephones respectively.

The Western Union Telegraph Company is the only domestic telegraph carrier. In addition to public message telegraph service, the carrier provides a teletype service (Telex) and several specialized information services. Western Union's revenues in 1967 totaled $336.8 million by comparison with AT&T's total revenues of $13,284 million.

Three companies are engaged in providing most overseas telegraph, teletype, and data services: RCA World Communications, a subsidiary of Radio Corporation of America; ITT World Communications, a subsidiary of International Telephone and Telegraph Company; and Western Union International (no corporate relations with Western Union). Operating revenues of the three companies in 1966 totaled $121.5 million. AT&T, GT&E (through its subsidiary, Hawaiian Telephone Company), and ITT (through its subsidiaries in Puerto Rico and the Virgin Islands) provide overseas telephone services. Revenues in 1966 totaled $145 million.

The Communications Satellite Corporation (Comsat) provides satellite circuits to the aforegoing overseas carriers. Comsat's operating revenues in 1967 totaled $18 million.

Some 650 radio common carriers (also called miscellaneous common carriers) and some 320 telephone companies offer land mobile radio telephone service to the public. In 1966 revenues of companies in this category totaled about $25 million.

Some companies which are engaged in noncommunication activities are capable of providing some or most of the intercommunication services which they need in order to carry on their principal business activities. We shall refer to those companies as "secondary intercommunication companies." Historically, railroads and pipelines have been the principal secondary intercommunication companies because they were able to string along their rights of way the wires required for telephone and telegraph operations. The advent of terrestrial microwave and satellite communications enables other companies to provide their own intercommunications. Therefore, an increasing number of companies may seek to become secondary intercommunication companies.

One of the principal obstacles which has been placed in the path of companies becoming secondary intercommunication companies has been the refusal on the part of the telephone companies in most instances to permit the former to interconnect their communication facilities with the nationwide telephone network. We shall see later how public policies which have sanctioned such refusals are being modified as a result of increased pressures exercised by existing and potential secondary intercommunication companies.

Mass Communication Companies

There are in excess of 6,500 AM and FM radio stations and over 1,000 television stations operating in the United States. The TV stations reach over 54 million homes. The mass communication companies which operate those stations range from the three large radio and television networks, the American Broadcasting Corporation, the Columbia Broadcasting System, and the National Broadcasting Corporation, through companies which own groups of stations up to the maximum number allowed by the Federal Communications Commission, down to companies which own a single station. Industry revenues in 1967 totaled in excess of $3.21 billion, of which television revenues constituted $2.3 billion and radio revenues $0.91 billion. Approximately 50 percent of television revenues were those of the three networks and their owned and operated stations.

In April 1969 there were in existence about 2,260 CATV systems serving about 3.6 million homes in 3,650 communities. Approximately an equal number of additional CATV franchises have been granted by local communities or facilities are already under construction. While some systems have more than 10,000 subscribers, the average system has about 1,800 subscribers. An increasing number of CATV systems are owned by companies which also own radio and television stations.

The following figures[5] which show the ratio of net value of plant to revenues help to underscore the basic difference between the intercommunication business and the mass communication business.

	Companies	
	Intercommunications	*Mass Communications*
Net Value of Plant	$33 billion	$0.827 billion
Revenues	13 billion	3.1 billion
Ratio	2.50 : 1.00	0.27 : 1.00

Principal Regulatory Agencies

As we have stated earlier, the terms "regulation," "regulate," and "regulatory" are used very broadly within the context of this study. Therefore, when we speak of regulatory agencies we have reference to all levels of government—

Federal, state, and local — and all branches of government — legislative, executive, and judicial — as well as the so-called independent regulatory commissions which play a role in determining the structure of the communications industry.

The principal Federal agencies which specialize in communications regulation are (1) the Federal Communications Commission, a seven-member independent regulatory commission vested by Congress with broad powers to regulate the industry, and (2) the Director of Telecommunications Management, to whom have been delegated regulatory powers granted to the President by several statutes dealing with communications.

Being concerned generally with monopoly and competition in our economy, the Department of Justice must be listed as one of the principal agencies which have an impact on the structure of the communications industry. Since antitrust cases prosecuted by the Department of Justice and private antitrust suits are decided by the Federal courts, these must also be listed among the principal regulatory agencies which may determine the structure of the communications industry.

Because they are the largest users of communication services, and because the exercise of their procurement powers affects the structure of the communications industry, the Department of Defense and the General Services Administration must also be included among the principal Federal agencies which regulate the industry structure. As stimulator of new communication technologies, the National Aeronautics and Space Administration (NASA) has an important potential in shaping the future structure of the industry.

Since the advent of satellites, Congress and the President have become increasingly concerned with distributing communication functions among different companies. Therefore, they must not be forgotten in listing government agencies which regulate the structure of the industry.

State and local regulation of communications can influence

importantly the structure of the industry. This has been brought home forcefully in recent years in the case of the regulation of cable television systems. On the state and local levels, the several branches of those governments interact in determining the structure of the industry.

Summary of Conclusions

Nature of Conflicts

It has been recognized for some time that the regulation of industry is a political process rather than one in which neutral technicians apply appropriate principles of law and economics.[6] If this is true with regard to industry regulation in general, it is true particularly with regard to the regulation of the structure of industries.

The conflicts that arise over the distribution of new business opportunities within and beyond the borders of the communications industry are polycentric conflicts in which numerous interested parties are warring with each other on several fronts.[7] Individual conflicts are interconnected like nodes in the web of a spider. It is difficult, if not impossible, to deal separately with such conflicts as various interested parties, tugging on different parts of the web, distort the positions of the nodes in relation to one another.

Chairman E. William Henry of the Federal Communications Commission, in stating his reasons for denying AT&T's petition for reconsideration of the Commission's order directing a formal investigation of AT&T's rates and earnings, spoke of the "extraordinary confluence of regulatory problems" which made the formal investigation necessary. While Chairman Henry thought that such confluence had occurred "seldom if ever before in the Commission's history," the truth is that the Commission ordinarily closes its eyes to the polycentric nature of the conflicts and treats them as readily separable from each other.[8]

The parties to such conflicts may be individual companies, industry segments, or entire industries. Different government agencies become involved in such conflicts either because the pursuit of new business opportunities requires governmental approval or because some of the parties appeal to the agencies for protection from what they consider unfair competitive practices engaged in by others. In such conflicts, therefore, we find competition for regulation, with each of the contending parties seeking to persuade government regulators to resolve conflicts in their favor.

No single government agency has the authority to decide such conflicts. Different agencies are likely to favor different solutions, and coordination among agencies is inadequate at present. The need for improved coordination can only be met on the macropolitical level of the Federal government.

Imprecise Yardsticks for Conflict Resolution

Institutional economists have acknowledged that the understanding of relationships between industry structure and industry performance is extremely limited even under relatively static technological conditions, and it is even more tenuous where such relationships are affected by rapidly changing technologies.[9] Therefore, present knowledge is inadequate to provide reasonably precise yardsticks with which to measure relationships between structure and performance of the communications industry *in general,* and to apply such yardsticks in selecting the best structure. It is even difficult to secure substantial agreement among government regulators on optimal solutions of particular, more limited conflicts over the distribution of communications functions. It is possible, however, to make informed judgments on a case-by-case basis as to which proposals for conflict resolution should be ruled out. Proposals should be rejected that are likely to impair the capacity within the system for reaching decisions regarding introductions of technological changes and thus to affect adversely the long-range performance of the system. This will leave a range of proposals from which to choose. Such an approach can facilitate greatly the difficult tasks faced by government regulators in adjudicating conflicts over the distribution of communications functions.

Need for Broad and Flexible Goals

Given the absence at present of reasonably precise knowledge of the relationships between industry structures and industry performance under conditions of rapidly changing technologies, it is desirable in the public interest as well as in the interest of the industry not to pursue definitive governmental policies but rather to reach for broad and flexible regulatory goals. Such goals should include two which stand in a relationship of tension to each other: (a) to preserve and to strengthen insofar as possible incentives and capacities of individual companies to plan their future roles in the communications system, and to take advantage of new technologies or meet new demands; and (b) to make the plans of individual companies compatible with a view to improving, or at least not affecting adversely, the future performance of the communications system as a whole. The achievement of these goals and the task of resolving conflicts over the ways in which these goals should be achieved in specific situations become more difficult as the number of companies and the points of view held by such companies increase.

To facilitate conflict resolutions, company managers must be prepared to participate actively in such efforts. Government regulators, on the other hand, must focus not only on questions of who is to produce what goods and services at a given time but also on the long-range impact of particular decisions on the incentives and capacities of the respective companies to be progressive and dynamic organizations.

Need for Appropriate Conceptual Framework

In order to achieve more effective and expeditious resolutions of polycentric politico-economic conflicts, industry and government must supplement formal regulatory processes, which tend to polarize positions taken by interested parties, with informal and accommodative interactional processes. Such processes require, first of all, a conceptual framework sufficiently comprehensive to include multidisciplinary reasoned arguments advanced by industry and government. The framework must accommodate in particular facts and arguments advanced in connection with questions of competition and monopoly. The framework must be appropriate for developing ranges of alternative feasible methods of policy coordination and conflict resolution. It must, therefore, tend to promote trade-offs among diverse factors instead of tending to polarize positions taken regarding them.

Need for Appropriate Organizations and Procedures

In addition to a conceptual framework, regulatory processes, in order to be appropriate for the coordination of policies and the resolution of polycentric politico-economic conflicts, require suitable government and industry organizations and procedures. These must be informal and flexible to promote the development of a range of feasible alternatives designed to provide specific communication services. In selecting one of the proposed coordinating methods or conflict resolutions as being in the public interest, extensive reliance will have to be placed on negotiation and bargaining. Negotiation and bargaining, however, tend to arouse suspicion because the fear of improper deals and conspiracies in restraint of trade is always present. It will be necessary, therefore, to surround negotiating procedures with appropriate safeguards, just as other regulatory procedures had to be "reformed" before they became accepted as legitimate government procedures.

Research Methods Employed

This study involved first and foremost an analysis of how the present structure of the communications industry evolved, with particular attention to the ways in which past conflicts over the distribution of new business opportunities have been resolved. Based on this analysis, I reached some tentative conclusions with regard to the adequacy of these conflict resolution processes. From these conclusions a number of tentative recommendations designed to strengthen these conflict resolution processes have been developed.

The analysis together with the tentative conclusions and recommendations were submitted in writing to a number of individuals in the communications industry and in government with a request for comments. The individuals were selected partly because of the positions which they hold and partly because they have been known to me over the years as individuals who hold independent views and who do not mind sharing such views on a confidential basis. These people then volunteered to express their opinions in writing or in personal interviews; in some instances they were willing to do both. Interviews were also conducted with individuals who have had considerable experience with structural regulation of other very large and complex systems, such as the transportation and energy systems.

Last but not least, I have sought to check my judgments and proposals against insights expressed by writers in such diverse fields as institutional economics, political science, systems analysis, organizational behavior, and individual behavior which seemed relevant to the problem areas dealt with in this study. Where such insights seemed to be particularly significant, reference is made in the text to the particular works of these authors.

PART I

The Evolution of the Present U.S. Communications System

CHAPTER II

Intercommunications

The Landline Companies

The core of the U.S. intercommunications system consists of facilities owned and operated by the so-called landline companies. These companies render domestic public telephone and telegraph services primarily but no longer entirely by landlines consisting of wires and cables. Increasingly radio microwave facilities have been included along with wires and cables for long-distance telephony and telegraphy. Furthermore, the capacity of the landlines has been augmented greatly through the use of coaxial cables and may soon be augmented further through the use of domestic satellites. The landline companies also use radio for public mobile telephone service — a service which is still in its infancy.

The relations between landline companies, their relations with their customers and their hardware suppliers, and the impact of the Federal government on all these relations have contributed greatly to the evolution of the present configuration of the U.S. communications industry.

Telegraph Operations

It was during the early years of the antitrust-reform movement under the Wilson Administration that the Federal government adopted the policy that the public should not be placed in a position where it would have to rely upon a single company for substantially all domestic communication services. In compliance with this policy, the American Telephone and Telegraph Company, in the so-called Kingsbury

Commitment,* undertook (1) to divest its holdings in Western Union which it had acquired in 1909; (2) to refrain from acquiring additional independent telephone companies except under certain limiting circumstances; and (3) to interconnect its facilities with those of the independents in order to enable the latter to offer to their customers nationwide telephone services.

As a result of the government's insistence upon competition, or at least multiple companies, and AT&T's compliance therewith, the system was divided for approximately two decades into telephone (voice) and telegraph (record) communications. AT&T and the independents offered the former while Western Union and Postal Telegraph Company furnished the latter. (The telephone companies, however, leased private lines to their customers for various services, including record services.)

This division at the behest of the Federal government constituted a re-establishment of a state of affairs which had prevailed between 1879 and 1909 as the result of an intercompany agreement between National Bell and Western Union. Pursuant to this agreement, the latter sold its telephone patents and properties to the former, recognized the priority of the Bell patents, and secured a commitment from Bell that the company would stay out of the telegraph field. The National Bell–Western Union agreement also terminated efforts of Western Union to enter the telephone field.*

In 1909 AT&T decided to enter the telegraph field by acquiring control of Western Union, then the dominant but

---

* Contained in a letter dated December 19, 1913, from N. C. Kingsbury, Vice President of AT&T to Attorney General McReynolds (later appointed by President Wilson an Associate Justice of the U.S. Supreme Court).

* When in 1876 Alexander Graham Bell obtained the first of several patents on his invention of the "talking machine," he offered the patent to Western Union for $100,000. The company refused the offer. Subsequently, realizing its mistake, Western Union attempted to enter the telephone field and for this purpose acquired patents which were competing with Bell's patents.

not the only telegraph company. Theodore N. Vail, the president of AT&T, gave the following reasons for the company's decision:

> . . . if Western Union were controlled by the telephone company, all of its lines could be utilized to a greater or lesser extent for toll lines and long distance telephone business. The telephone company will be obligated to spend a great many millions of dollars, fully as many as the telegraph company will cost, to provide toll line facilities which could be largely avoided if it had the use of Western Union facilities, or the control rather — as the mere use, without the absolute control, would be of no account.[1]

The combined telephone-telegraph operations were short lived. Following the consolidation, the Department of Justice received complaints from Clarence Mackay's Postal Telegraph-Cable System, a competitor of Western Union, and from independent telephone equipment manufacturers concerning AT&T's telephone and telegraph operations. An investigation was conducted at the request of the Department by the Interstate Commerce Commission, which in 1910 had been given authority by Congress to regulate interstate communications companies. Following this investigation, and upon the insistence of the Department, a division of fields between voice and record communications took place which was to last until after World War II.

Following World War II, the market for communication services changed greatly.* The demand for telephone services

---

* The information contained in this section is based largely on the *Report of the Telephone and Telegraph Committees of the Federal Communications Commission in the Domestic Telegraph Investigation,* Docket No. 14650, April 29, 1966. The investigation was instituted by the Commission in 1962 for the purpose of examining into the problems related to the decline of the message telegraph service and possible alternative solutions of these problems. The Committees consisted of Rosel H. Hyde, Chairman, and Commissioners Robert T. Bartley and Kenneth A. Cox. The Committee's Report is based largely on an earlier staff report in this investigation.

rose steeply while the demand for public message telegraph services declined sharply. In addition, demands increased for new types of communication services which had been developed to meet the needs of far-flung business firms: teletypewriter exchange, alternate voice/record, and voice/data services.

These technological advances had two important consequences: (1) Western Union absorbed the ailing Postal Telegraph Company, a subsidiary of ITT, in accordance with legislation specifically authorizing such consolidation; and (2) Western Union attempted to diversify its public message telegraph business by offering teletypewriter exchange (Telex) and alternate voice/record and voice/data private line services. These offerings brought the company into direct competition with AT&T's teletypewriter exchange services (TWX).

One of the complicating factors in this competitive relationship between AT&T and Western Union was the circumstance that following World War II Western Union began to utilize intercity and intracity circuits leased from AT&T.

Until World War II Western Union had largely relied on its own *intercity* telegraph channels to render telegraph services. Following World War II the company increasingly replaced its obsolete open wire facilities with circuitry leased from the Bell System. By 1963 the ratio of Western Union leased to owned facilities stood at 70:30. With the increasing reliance since then on Western Union's own transcontinental microwave system, the ratio has become reversed and in 1967 stood at 20:80.

The dependence of Western Union on AT&T with regard to *intracity* circuits has continued to remain very great. In 1963, 80% of the company's intracity requirements were leased from the Bell System. This figure has remained substantially unchanged. Additionally, Western Union is leasing facilities from independent telephone companies.

The availability of these facilities has enabled Western Union to provide private line and exchange-type switched services without making large-scale investments in transmission facilities of its own to render these services.

AT&T has permitted Western Union to use the leased circuits freely to render telegraph services. However, in order to protect the exclusiveness of its voice communication business, AT&T has required Western Union to use AT&T's telephone instruments at the terminals of the leased circuits when offering alternate voice/record or voice/data services.[2]

Western Union has also attempted to compete with AT&T in rendering television transmission services. This field had been the exclusive business of the Bell System. When Western Union had constructed a portion of its microwave system it sought to enter the television transmission services, and toward this end it sought an interconnection of its microwave system with Bell's television transmission facilities. This request for interconnection was refused and Bell's refusal was upheld by the Commission.

The FCC's Telephone and Telegraph Committees in their report of April 29, 1966, in the Domestic Telegraph Investigation recommended certain steps designed to make Western Union a "viable entity." These steps included transferring Bell System's teletypewriter exchange service (TWX) to Western Union, precluding AT&T's re-entrance into the exchange telegraph market by making an exchange teletypewriter offering available over the toll telephone network; elimination of interconnection restrictions that prevented Western Union from competing effectively with the Bell System in the provision of private line services; and viewing interconnection between Western Union and Bell facilities as a means of promoting competition in the area of television program transmission service.

The sale to Western Union by AT&T of the latter's TWX facilities was finally agreed to by the two companies on January 15, 1969. The negotiations leading up to the sales agreement had extended over a quarter of a century. The sale was viewed by the Congress in 1943 as a desirable way of restructuring the domestic intercommunication system. When Congress in that year enacted the Domestic Merger Act, which authorized Western Union's acquisition of the Postal Telegraph Company, it also authorized the merged company "to acquire the domestic telegraph facilities . . . of any carrier which is not primarily a telegraph carrier." This general language was designed not only to authorize Western Union's acquisition of Bell's TWX facilities; it was also intended to nudge AT&T into entering into negotiation with Western Union for the sale of the TWX facilities.

The negotiations opened that same year and continued for a period of two years. Late in 1945, however, Western Union's financial ability to acquire the facilities was impaired by reason of retroactive wage increases ordered by the War Labor Board. That order practically wiped out the company's financial reserves and materially increased its future operating costs.

The negotiations were resumed in 1964 but were interrupted when Western Union in the Commission's telegraph hearings suggested that the sales of the facilities would constitute a recognition by AT&T of the principle of voice/record separation. In a letter dated November 19, 1964, from Commissioner Rosel L. Hyde, Chairman of the FCC Telephone and Telegraph Committees, to Frederick R. Kappel, Chairman of the Board, AT&T, the Commission assured AT&T that it would not construe the sale as acquiescence by AT&T in a program calling for complete separation of voice and record services. Thereupon, the negotiations were resumed. The FCC's Telephone and Telegraph Committees subsequently stated in their report of April 29, 1966, that at that time a policy of domestic voice/record separation was not justified and that such a policy might not serve the public interest.

The negotiations were interrupted again in mid-1968 when Western Union began merger discussions with Computer Sciences Corporation of Los Angeles. The suspended negotiations were resumed after the merger discussions were terminated.

In the final sales agreement between AT&T and Western Union the latter company warrants that no merger agreement or public tender offer is outstanding and that there are no merger or consolidation proposals in existence. Substantially the same conditions must prevail at the time of closing the sale.

The sales agreement is subject to regulatory approval which is expected to be forthcoming since the transfer of the TWX facilities was recommended by the FCC as a step in creating an integrated record message system.

Telephone Operations

When the Kingsbury Commitment was made by AT&T in 1913 there were in existence some 20,000 independent telephone companies serving about 3.6 million independent telephones, in comparison with Bell's about 5.1 million telephones.[3] In 1968 there were about 87 million Bell telephones and some 2,100 independent telephone companies operating about 17.2 million telephones.[4] Mergers and consolidations have led to the reduction in the number of independent companies. Among the independents, four major companies have emerged: General Telephone and Electronics Company, United Utilities, Inc., Continental Telephone Corporation, and Central Telephone and Utilities Corporation. In 1967, these four companies served about 11.96 million telephones: GT&E about 8.7 million; United about 1.5 million; Continental about 1 million; and Central about 0.76 million telephones.

The independent telephone companies serve geographical areas different from those served by the Bell System. Like the telephone companies affiliated with the Bell System, the independent companies are monopolies within their respective areas. On the basis of AT&T's Kingsbury Commitment, they are assured of interconnections with the Bell System, and by Federal law they are protected from the threat of being absorbed by the Bell System.[5]

Unlike the Bell System, however, the independent companies are not subject to certain constraints which were imposed on the Bell System through a consent decree which was entered in 1956 in an antitrust suit brought in 1949 by the Department of Justice. The details of that decree are discussed in the next section. At this point, however, it should be noted that the independent companies are free to merge with each other. They may engage in business activities which are not common carrier communication activities; they may manufacture communications equipment for their own telephone operations and for sale to other companies and the general public.

The capacity on the part of the independent companies to undertake these various activities has enabled these companies to play roles within their respective geographical areas which differ somewhat from the roles played by the telephone companies affiliated with the Bell System. Some of these differences will be discussed subsequently because they are likely to have an important impact on the future structure of the communications industry.

Communications Hardware

General Telephone and Electronics has owned two manufacturing subsidiaries (Automatic Electric and Lenkurt Electric) for quite some time and it acquired Sylvania in 1959. Continental and United have acquired manufacturing subsidiaries only in recent years. These subsidiaries manufacture communications hardware for sale not only to the operating telephone companies within their respective systems but also to other firms and governments in the United States and abroad.

In owning their own manufacturing subsidiaries, three of the large independent telephone companies have followed in the footsteps of AT&T, which owns its own manufacturing and supply unit, the Western Electric Company. AT&T's ownership of Western Electric, and the ownership by other telephone companies of their own manufacturing and supply units constitute an important facet of the present configuration of the U.S. communications industry. This facet has resulted in several confrontations among communications firms and between such firms and government agencies.

In January 1949 the Department of Justice brought an antitrust suit against AT&T seeking to force its divestiture of Western Electric and to split Western Electric into three separate companies. The suit was terminated seven years later by a consent decree. The decree permitted AT&T to retain its ownership of Western Electric but precluded Western Electric from manufacturing any equipment of a type which was not sold or leased to Bell System companies for use in furnishing common carrier communication services.

The consent decree has been criticized on a number of occasions as not having achieved the purposes of the 1949 suit. In 1961, for example, the Chairman and members of the Antitrust Subcommittee of the Judiciary Committee of the House of Representatives questioned representatives of the Department of Justice and the Federal Communications Commission with regard to the effectiveness of the consent decree in achieving the objectives of the antitrust laws. A majority of the Subcommittee urged a re-evaluation of all the facts and circumstances and appropriate action by the Department.

In 1964 the Department of Justice filed an antitrust suit

against General Telephone and Electronics to block the acquisition by GT&E of several independent telephone companies on the West Coast.[6] The Department contended that the acquisition would tend to foreclose the communications equipment market to independent manufacturers. This suit was subsequently withdrawn by the Department on the grounds that it was incompatible with the AT&T consent decree. In October 1967, however, a private antitrust suit against GT&E and the Hawaiian Telephone Company which was acquired by GT&E was filed by ITT. This suit, which is based on substantially the same ground as the government suit, was still pending in the courts when this study was made.

Operating telephone companies constitute one part of the market for communications hardware. Another part of that market is represented by over 100 million telephone subscribers. Traditionally, AT&T and the independent telephone companies have preempted that market by requiring their subscribers to use almost entirely telephone equipment and other related communications hardware owned and furnished by the companies.

The companies accomplished this by filing with the state regulatory commissions and the FCC tariffs which, with rare exceptions, precluded subscribers from owning and attaching their own equipment. Such attachments were referred to as foreign attachments.

The policies against foreign attachments which were followed by the telephone companies over the years were challenged in the so-called Carterfone proceeding.[7] That proceeding started out in 1966 as a private antitrust suit brought by Carter Electronics Corporation, a small Dallas-based manufacturing company, against AT&T and GT&E in the U.S. Federal District Court in Texas. The climax came in 1968 in a major FCC proceeding in which the Commission held that the interstate tariff provisions of the two telephone companies prohibiting foreign attachments were unreasonable, unlawful, and unreasonably discriminatory, and ordered the provisions stricken from the tariffs.[8]

Carter's original objective was quite limited. He sought a continuing market for his Carterfone: a cradle-like instrument which utilizes an acoustic/inductive mechanism to interconnect private radio communication systems with the telephone companies' public telephone network by placing a conventional telephone headset in the instrument. The Carterfone was particularly valuable to petroleum pipeline and other companies which were operating private communication systems, since the interconnection of their private systems with the public network enhanced their usefulness. Between 1959 and 1966 Carter sold about 3,500 Carterfones in the Unitcd States and overseas. The refusal of the telephone companies to permit the use of the Carterfone by telephone company customers destroyed the market for the product. Carter brought suit against the telephone companies on the grounds that the refusal was unlawful under the antitrust laws.

The court referred to the FCC the regulatory issue, namely, the question whether the foreign attachment prohibitions of the telephone companies were applicable to the Carterfone. Carter sought to prove that the tariffs did not apply to the Carterfone since that particular piece of communications equipment did not have an adverse physical effect upon the operation of the telephone system. AT&T and GT&E maintained that it did have such an effect.

Carter's position was supported by the American Petroleum Institute, representing the petroleum companies which had sought to use the Carterfone, and by the National Retail Merchants Association, representing department stores and other retail establishments which sought to interconnect their private communication systems with the public network.

Most important, however, the Department of Justice intervened and contended that the tariff provisions against foreign attachments, while of long standing, had growing anticompetitive consequences which were contrary to the policies underlying the antitrust laws. The Department sought the cancellation of the present foreign attachment provisions and the substitution of provisions which would affirmatively permit use of such foreign attachments, subject only to such reasonable technical specifications as the Commission might find essential to maintaining the technical integrity of the carriers' systems and the ability of the carriers to fulfill their obligations.

The Commission, in its unanimous decision of June 26, 1968, held not only that the telephone companies would be required to change their foreign attachment provisions in the future, but also that those provisions should be considered unlawful retroactively to 1957, the date when the tariffs were filed. The companies sought FCC reconsideration of the decision but to no avail. They then sought judicial review of the Commission decision but later changed their minds and withdrew their petitions.

On September 13, 1968, the day following the FCC's refusal to modify its decision, AT&T filed new tariff provisions which were designed to permit foreign attachments and interconnections of private systems subject to specified limitations. Several government agencies, trade associations, and individual companies intervened in the proceeding for the purpose of securing the rejection of the new provisions by the FCC on the grounds that the limitations did not comply with the Commission's decision and order. Large users of communications and manufacturers of communications hardware were represented by the opposing associations.

Informal negotiations ensued between AT&T and the Commission staff, and on October 22, 1968, AT&T submitted modified tariff provisions with regard to attachments and interconnections. The Commission permitted the modified tariffs to go into effect as of January 1, 1969. It also held that the telephone sets supplied by the telephone company are integral parts of the telephone network.

The Commission decided against holding formal proceedings with regard to the new tariffs, but instead it provided for a series of informal conferences at which interested persons and organizations including government agencies, user groups, and manufacturers could discuss what further changes might be necessary, desirable, and technically feasible.

THE LANDLINE COMPANIES AND RECENT RADIO TECHNOLOGIES

The development of microwave technology which began in the 1940s and regulatory decisions made in the light of that development have led to the establishment of private and public point-to-point microwave systems independent of the traditional landline companies. Microwaves are very short waves which utilize the upper regions of the radio spectrum above 890 megacycles. They are particularly well suited for high capacity point-to-point communications. Over longer distances such communications are carried on by means of relay stations in line of sight of each other. Distances vary from 25 to 50 miles depending on the intervening terrain.

The private* systems are owned and operated by government agencies or business firms which require communication services on a large scale in support of other activities. The public systems offer specialized common carrier communication services to the public. Such services may or may not be in competition with similar services offered by landline companies. The growing importance of the private and public microwave systems has led to a number of conflicts

---

* "Private" in this context is used in contrast to "common carrier." Thus a private system may be operated, for example, by an agency of the Federal government, a state, a municipality, or a business firm.

between the landline companies and parties interested in promoting the use of such systems.

Point-to-Point Microwave: Private Systems

Private point-to-point microwave systems have emerged as an alternative method of providing needed communication services which otherwise would have been provided by the landline companies. Prior to the advent of microwave, only the so-called right-of-way companies (i.e., railroads, pipelines, and electric power companies) were capable of operating private communications systems, since they could string the necessary wires along their rights-of-way. Today, many of the right-of-way companies are substituting microwave for wire communication. In addition, companies which do not have a right-of-way, as well as states and local governments, may use microwave facilities to provide their own long-distance communication services.

Microwave communications can handle telegraph, telephone, teletype, facsimile, digital data, and TV relay transmission, as well as "pushbutton" remote control of machines and other devices used throughout industrial and other business systems. "Microwave has become radio's jack-of-all-trades." [9]

The regulatory decision which paved the way for the extensive development of private point-to-point microwave systems was reached by the Federal Communications Commission in 1959 in a proceeding entitled, "In the Matter of Allocation of Frequencies in the Bands over 890 Mc." (Docket No. 11866). In this proceeding the Commission addressed itself to two broad issues: (1) the adequacy of the supply of microwave frequencies and the terms on which such frequencies should be made available to private systems, including the sharing of such systems by private users and interconnections between private systems and common carriers; and (2) the economic impact of private systems on common carriers, the users of common carrier services, and the public.

Heretofore, it had been the general policy of the Commission to make microwave frequencies available for private systems (with the exception of right-of-way companies) only on an experimental rather than a permanent basis.

Telephone and telegraph common carriers generally opposed the allocation of microwave frequencies to private systems (other than public safety organizations and so-called right-of-way companies) on several grounds: namely, that insufficient frequencies were available; that common carriers could make the most efficient use of available frequencies; and that private systems would engage in "cream-skimming" which would tend to undermine the economic base of the common carrier systems. The private users, actual as well as potential, and the manufacturers of communications hardware argued for liberal rules of eligibility and freedom of choice for users to establish their own systems or to obtain facilities or services from the common carriers.

The decision reached by the Commission supported the latter viewpoint.[10] The Commission found that adequate frequencies above 890 Mc. were available to take care of present and reasonably foreseeable future needs of both common carriers and private users for point-to-point communications systems. The Commission, however, denied, with certain exceptions, general permission to private users to share frequencies on a cooperative, nonprofit, cost-sharing basis. It did so on the grounds that such cooperative facilities might lead to undesirable situations where the cooperatives might have many of the attributes of communications common carriers. Subsequently, however, the Commission modified this policy and it is now permitting extensive sharing of private systems. Interest in shared systems, however, has not been as vigorous as had been anticipated.

With regard to interconnections between private systems and common carriers, the Commission thought it inappropriate to establish standards and criteria applicable to all situations involving interconnections, since these were matters governed by tariff regulations and practices which are required to meet the statutory standard of justness and reasonableness. Such tests, the Commission stated, can only be applied on the basis of the facts involved in a particular case and in accordance with specific procedures established for the regulatory administration and review of tariffs.

As to any possible adverse economic impact on the common carriers and the users of carrier services, the Commission found that under the provisions of the Communications Act of 1934 it had no express obligation to extend such protection. Additionally, however, the Commission found that there were sufficient facts in the record which tended to indicate that no substantial adverse economic effects would result from the licensing of private point-to-point systems.

On the other hand, the Commission felt that the liberalized availability of microwave to private systems would afford a competitive spur in the manufacturing of equipment and in the development of the communications art. This point had been stressed particularly by the Department of Justice in a letter to the Commission. In its Report and Order of September 28, 1960, upon reconsideration the Commission stressed that "the 'competitive spur' was only one of several factors considered by the Commission in its determination, in the public interest, to liberalize its licensing policy to permit the establishment of private microwave systems."

Under the liberalized licensing policy the Commission had issued, as of January 1, 1969, 7,600 microwave station authorizations to some 810 licensees. These stations provide about 167,000 miles of microwave paths.[11]

The largest number of these authorizations are held by right-of-way companies (petroleum about 2,000; power about 1,900; and railroads about 850). Public safety licensees constitute the next highest category (state and local governments about 900; police about 250; highway maintenance about 225; forestry-conservation about 65; and fire about 35). The third class is made up of industrial radio service licensees who hold about 630 authorizations.

The sizes of the systems in each of the three categories differ widely. Among some of the largest systems are those of the Southern Railway System, the Pacific Gas and Electric Company, and the State of California. The average system has 9.5 microwave stations.

Many of the systems combine point-to-point microwave operations with mobile communications, thus making possible within large geographical areas communications both between moving vehicles and between fixed points.

The question of the extent to which such systems may be interconnected with the communication network owned by the telephone companies was reviewed by the Commission in the Carterfone proceeding. As discussed above, the Commission held that tariff provisions which limited foreign attachments of equipment and interconnections of private systems were illegal, and ordered the telephone companies to submit new tariffs. The liberalized interconnection privileges for private systems are likely to provide a substantial additional stimulus to the establishment of new private systems. To what extent the telephone companies will be able to offer facilities and services at rates designed to meet the competition of private systems depends on the final outcome of other Commission proceedings in which the level of those rates is at issue.

Point-to-Point Microwave: Public Systems

Point-to-point microwave systems may be used not only by landline companies to supplement their cable facilities and by companies or government entities which use such systems to supply their own communications services; such systems may also be used — with FCC authorization — by special service common carriers to offer limited communication services in particular geographical areas or for specified purposes. A number of such carriers have been established for the primary purpose of serving CATV systems. Such carriers are engaged in the transmission of television programs from distant television stations to the CATV systems, enabling these systems to carry additional television programs. Other carriers provide links between television studios and transmitters. Such carriers may also provide services linking mobile television pick-up trucks with studios or transmitters.

In isolated instances, proposals have been submitted to the FCC for the establishment of carriers which are designed to serve a greater diversity of customers. Microwave Communications, Inc., for example, has made application to the Federal Communications Commission for permission to construct and operate a microwave system between Chicago and St. Louis. If the application is granted, the company would seek interconnections between its system and the systems of the landline companies.

The company proposes to serve such customers as time-sharing computer service organizations, management information systems users, barge lines and other transportation companies, as well as colleges and other educational institutions. Such customers have contended that they require communications services in large quantities at rates lower than those at which comparable services are offered by the landline companies. Also, these customers seek to attach communications and other equipment of their own choice instead of equipment furnished by the landline companies. Before the Carterfone decision required the landline companies to file modified tariffs with regard to foreign attachments, these companies refused to permit their customers to attach customer-owned equipment to the switched dial telephone network. This made it necessary for customers either to lease private lines from the landline companies or to use equipment furnished by the landline companies.

Independent manufacturers of communications hardware who are seeking access to markets for their equipment favor entry of special service common carriers since such entry would broaden the markets for communications hardware.

The landline companies opposed Microwave Communications' application on technical as well as policy grounds. They contended that entry into business by special service common carriers would lead to cream-skimming by such carriers and would undermine the landline companies' rate structures which are based on averaging rates for high and low density traffic routes. For instance, the manager of the Eastern Oregon Telephone Company, Walter Karnopp, warned that the Microwave Communications proposal was an "invitation to disaster." He stated that the telephone industry "cannot lose the only declining unit cost business it has, the high usage toll routes, without destroying the entire average rate-making structures that have enabled the industry to bring subsidized exchange service to millions who would otherwise be unable to afford a telephone." [12]

The Commission handed down its decision in the six-year-old proceedings on August 14, 1969, granting Microwave's application. Assuming that the 4 to 3 decision stands,* it will significantly affect the future structure of the U.S. communications industry by permitting special carriers to offer services over high-volume, low-cost routes, thus weakening the long-established practice of average national rate making. Following the decision Microwave applied for authority to operate a similar system between Chicago and New York. Another company, Interdata Communications, Inc., had already applied for authority to establish a system between New York, Philadelphia, and Washington.

---

* The members split along party lines which is rare in the case of the FCC.

Unless the regular carriers are willing to provide local loop interconnections between Microwave's system and its customers in Chicago and St. Louis, the FCC will re-open the proceeding for the purpose of determining whether the regular carriers should be required to do so.

The dissenting commissioners stressed the desirability, for social as well as other reasons, of maintaining nationwide as well as state cost averaging, and on those grounds alone would have rejected Microwave's application. The majority felt that Microwave should be given an opportunity to show that it can compete productively by better meeting the needs of potential subscribers for low-cost communication services.

Land Mobile Radio: Private Systems

Private* land mobile radio systems are very numerous. As of June 30, 1968, they had a total of some 2.66 million transmitters.[13] The individual systems vary greatly in size and not a few of them are operated, as has been pointed out above, in conjunction with private point-to-point microwave systems as, for example, in the case of several systems owned by railroads and pipelines.

The systems are used in connection with a large variety of operations: for example, police and fire; dispatching emergency crews for public utilities; railroad, truck, bus, and taxi operations; remote control of machinery and industrial processes; paging services for personnel in industrial establishments.

The FCC has expressed the view that the major problem facing these systems in large urban areas is the overcrowding of the limited spectrum space which the FCC in 1949 made available for these systems.[14] The Commission is now exploring alternative ways of remedying that situation, and it

* Here again, "private" in this context is used in contrast to "common carrier." Thus a private system may be operated, for example, by an agency of the Federal government, a state, a municipality, or a business firm.

has commissioned the Stanford Research Institute to study certain aspects of the problem. One of the alternatives under consideration is the re-allocation or sharing of frequencies presently allocated to UHF television. This proposal has brought mobile communications into conflict with broadcasters. The nature and resolution of this and other conflicts over the allocation of radio frequencies are discussed below.

Land Mobile Radio: Public Systems

Land mobile services are offered by radio common carriers and telephone companies. In making frequencies available for these services, the FCC stated in 1949:

> . . . we have taken particular care to provide a family of frequencies within which the development of common carrier mobile radio systems by enterprises other than existing telephone companies may take place. These dispositions have been effected advisedly, and with the purpose, among others, of fostering the development of competing systems, techniques, and equipments.[15]

At present the telephone companies operate about 1,500 base stations serving approximately 33,500 mobile units. About two-thirds of the base stations are operated by AT&T and one-third by independent telephone companies. The radio common carriers operate approximately 900 base stations serving about 28,000 mobile units.*

There are about 650 radio common carrier firms. These small business firms for the most part are only secondarily in the business of providing mobile radio services. Their principal business activities generally consist of providing telephone answering services or selling, leasing, and maintaining electronic communications equipment.

---

* These figures were furnished by the FCC. The National Association of Radiotelephone Systems claims that the number of base stations and mobile units served by the radio common carriers is somewhat higher.

The two-way voice communications services furnished by the carriers may be (1) message relay services, where an operator holds and forwards messages to or from mobile units; or (2) dispatch repeater services, which permit one vehicle to communicate with another vehicle without an operator as an intermediary; or (3) interconnected services, which permit a party in a mobile unit to speak directly to a party on the landline service or vice versa. The telephone companies usually offer only the last type of service.

Both radio common carriers and telephone companies are engaged in rendering one-way signaling services to individuals carrying appropriate receiving equipment on their persons.

In dealing with the FCC and other public agencies, as well as telephone companies, the radio common carriers are represented by a nationwide trade association, the National Association of Radiotelephone Systems. In the early 1960s the Association agreed with AT&T on a statement of principles and a model draft interconnection agreement. These serve as a general guide for Bell System operating companies and individual radio common carriers which are seeking interconnections with the former.

The growth of public land mobile two-way voice communications has been retarded severely by the unavailability of a sufficient number of radio frequencies in the 20 top urbanized areas of the nation where the demand for such services is greatest. In other areas, the demand for these services has been below available supply, with the result that not all the mobile units authorized by the Commission are in actual use.

One-way signaling services are considerably less expensive than two-way voice communications. Not only is the subscriber equipment inexpensive, but a radio channel which can accommodate about 100 two-way voice communication subscribers can serve many thousands of one-way signaling subscribers. Therefore, the Association expects this service to grow much faster in the near future than two-way voice communications.

In a FCC proceeding (Docket No. 16778), the Association sought the allocation of two radio channels for exclusive use by the radio common carriers for one-way signaling and opposed the allocation by the Commission of two radio channels for exclusive use by the telephone companies in providing one-way signaling services. The Association argued that one-way signaling was a message forwarding service which constitutes a new business for wireline carriers which historically have provided two-way services. Furthermore, it contended that radio common carriers will not be able to compete in the market place with the telephone companies in furnishing one-way signaling services. The telephone companies, on the other hand, claimed that the signaling device constitutes a simple extension of the receiver bell on the telephone instrument, and that the provision by the telephone companies of this service would not undermine the competitive ability of the radio common carriers to render this service.

The Commission decided to make available to the radio common carriers and the telephone companies separate radio channels to be used for one-way signaling. In making available such channels to the telephone companies the FCC imposed certain conditions with regard to the pricing of interconnecting telephone services which the telephone companies make available to the radio common carriers. The conditions are designed to assure that the radio common carriers can compete effectively with the telephone companies in offering the signaling services to the public. Judicial review of the Commission decision has been sought by Radio Relay Corporation, a New York common carrier paging operator, on the grounds that the Commission in making available frequencies to the telephone companies failed to take into account that the telephone companies would have "monopolistic advantages" in rendering paging services.[16]

The Landline Companies and U.S. Overseas Intercommunications

The earliest U.S. overseas intercommunication services were telegraph services, provided by means of undersea telegraph cables. These services were instituted in 1866. In 1901 low frequency radio began to be used in addition to undersea cables to provide overseas radio telegraph services. In 1927 high frequency radio made possible both overseas telegraph and telephone services. Undersea telephone cables for these services did not come into use until almost thirty years later. Finally, satellites began to be used in 1962 to furnish all kinds of overseas intercommunication services, including telegraph, telephone, data, and television relay services.

While undersea telegraph cables have gradually been abandoned, high frequency radio, undersea telephone cables, and satellites at present furnish a variety of U.S. overseas intercommunication services.

These technical developments, the initiative taken by different companies to exploit the developments commercially, and public policies, including particularly those designed to further rapid expansion during early development stages, have resulted in the present configuration of the U.S. communications industry with respect to overseas intercommunications.

When high frequency radio technology was first applied to provide overseas telegraph services, companies interested in its commercial exploitation and government regulators intent on promoting its rapid technical development persuaded the U.S. Congress that the new technology should be permitted to compete effectively with the older telegraph cable technology. Thereupon, Congress in the Radio Act of

1927 prohibited mergers of carriers by radio with carriers by cable if the purpose or effect of such mergers was substantially to lessen competition. This prohibition was designed to protect the development of the new technology which required less capital from being slowed down by the older cable technology which required larger capital investments.

The prohibition of mergers was re-enacted in 1934 as section 314 of the Communications Act of 1934. Today, however, that prohibition is no longer meaningful since all the major overseas telegraph carriers utilize all three modes of overseas communications: cables, radio, and satellites. There are at present three such major carriers: ITT World Communications, a subsidiary of ITT; RCA Communications, a subsidiary of RCA; and Western Union International, a company organized in 1963, which is not connected with Western Union but took over Western Union's overseas telegraph operations.*

None of the three international telegraph carriers is engaged in domestic telegraph operations, and Western Union, the only domestic telegraph carrier, is no longer engaged in overseas telegraph operations. This separation of overseas and domestic telegraph operations came about as a result of the domestic telegraph merger legislation enacted by the Congress in 1943. That legislation authorized Western Union to take over the ailing Postal Telegraph Company. One of the conditions of the take-over was the divestment** by

* Services to two geographical areas are rendered by Tropical Radio Telegraph Company, a subsidiary of the United Fruit Company which provides services between the United States and some Central American countries, and by United States-Liberia Radio Corporation, a subsidiary of Firestone Tire and Rubber Company, which renders services between the United States and Liberia. On May 8, 1968, RCA Communications and Tropical agreed to merge. The merger is subject to FCC approval. Western Union International has opposed the merger on antitrust grounds.

** The divestment was not effectuated, however, until 1963 when the facilities were transferred to Western Union International.

Western Union of its overseas telegraph facilities and operations.[17] Congress also forbade any merger between the newly merged domestic carrier and any international telegraph carrier.[18]

One of the important considerations which led to the adoption of those public policies was the apprehension on the part of the international carriers that the newly merged domestic carrier was likely to discriminate against them by transmitting over its own overseas facilities (if it were permitted to own and operate such facilities) those messages from points in the United States to points overseas where the senders did not specify, as many senders would not, a particular overseas carrier.

While there is thus a division between overseas and domestic telegraph operations, there is no such division between overseas and domestic telephone operations. AT&T is the principal domestic and overseas telephone carrier. GT&E by reason of its ownership of Hawaiian Telephone Company operates telephone cables to Hawaii, and ITT conducts telephone operations in Puerto Rico and the Virgin Islands.* When in 1928 AT&T sought to use high frequency radio for overseas telephone operations, there was no significant opposition to the Federal Radio Commission's (the predecessor of the FCC) granting the necessary licenses, nor in 1956, when the company sought a cable landing license for its first transoceanic telephone cable. Today AT&T utilizes all available modes of overseas communications, i.e., radio and cable facilities as well as satellite circuits provided by Comsat.

In 1964, however, in the so-called TAT-4 decision[19] which involved the granting of a cable landing license for the fourth transatlantic telephone cable, the Commission limited AT&T's overseas telephone business by denying the company

---

* In June 1969 RCA Communications acquired the Alaskan Communications System from the U.S. Air Force and became a domestic telephone operator.

the right to render overseas alternate voice/data services and reserving this portion of the overseas business exclusively for the U.S. overseas telegraph carriers. In placing this limitation on AT&T's operation, the Commission accepted the plea of the overseas telegraph carriers that they were in no position to compete adequately with AT&T with regard to these services in view of AT&T's vast domestic telephone network and the size of its sales force.

In the same decision the Commission ordered the joint ownership of the new TAT-4 transatlantic cable by all U.S. overseas telegraph and telephone carriers which desired to participate in such ownership. The division of such ownership was to be proportionate to present and reasonable foreseeable future requirements.

The telegraph carriers sought this ownership because their cable investments could be included in the rate base on which the companies' rates and earnings are computed. Rental payments were not so included, and previously the telegraph carriers had leased cable circuits from AT&T.

The advent of the communications satellite technology, the aspirations of individual companies in exploiting it, and public policies including particularly those aimed at furthering it, brought about important changes in the structure of the U.S. communications system with regard to overseas communications. (The potential impact of the new technology on domestic communications is discussed later on.)

During the earlier stages of commercial satellite development for overseas communications, AT&T, in putting up "Telstar" and planning for future satellites, sought to go it alone. In doing so the company was following a pattern which it had followed in utilizing high frequency radio circuits and undersea cables to furnish overseas telephone services. The earlier plans were abandoned, however, and an ad hoc committee of U.S. international carriers appointed by the FCC developed a substitute plan calling for joint ownership of satellites by AT&T and the U.S. overseas record carriers. Substantial opposition to this plan developed on the part of interest groups and proponents of various ideologies in and out of government. The aerospace companies opposed the plan because they feared that the proposed joint venture group would procure satellite hardware from AT&T, RCA, and ITT, all of which are electronic hardware manufacturers. This would enable the latter companies to expand their technical capabilities in the field of satellite hardware to the detriment of the aerospace companies which sought to maintain and enlarge their own capabilities in that respect.

These conflicting industry pressures contributed materially to the involvement of President Kennedy and the Department of Justice as well as Congress in the issue of satellite ownership and operations. Ultimately, the joint ownership plan was abandoned in favor of compromise legislation which created Comsat. Originally, Comsat was owned 50 percent by domestic and U.S. overseas communications common carriers and 50 percent by the general public. Subsequently, some of the carriers, and particularly ITT, sold substantial portions of their Comsat stock. As of February 1969 the carriers' holdings had fallen to 38 percent.

The Board of Directors of Comsat includes representatives of the carriers, of the public shareholders, and three public members who are appointed by the President of the United States with the advice and consent of the Senate. Originally, the carriers had six board members. Subsequently, in line with the reduction of carrier-held shares, their representation was reduced to four members.

The legislation which created Comsat circumscribed Comsat's role as follows: Comsat would be the sole chosen instrument through which the United States would participate in a commercial satellite system which would constitute part of an improved global communications network.[20] On the other hand — and this is equally noteworthy — while the legislation authorized Comsat to perform this particular function, it did not explicitly authorize Comsat to engage either in any other communications activity or in any activity outside the communications field. The legislation did not clearly determine whether Comsat would be authorized to use satellites for domestic communication services. Sec. 102 (d) of the Communications Satellite Act reads:

> It is not the intent of Congress by this Act to preclude the use of the communications satellite system for domestic communication services where consistent with the provisions of the Act nor to preclude the creation of additional communications satellite systems, if required to meet unique governmental needs or if otherwise required in the national interest.

The aerospace companies and other nonintegrated hardware manufacturers insisted that they should be assured of nondiscriminatory treatment by Comsat and the carriers with regard to purchase of satellite hardware. This resulted in the inclusion in the legislation of provisions calling for effective competition among the suppliers of hardware and services for the operation of the space segment and the ground stations. In order to assure fair competition in the procurement of satellite hardware, Comsat as well as the carriers are subject to the regulatory control of the Commission in that respect.

There was considerable controversy while the legislation was being considered over who should own ground stations — the carriers or Comsat. The legislation delegated to the Commission the decision who should own and operate ground stations in the United States:

> . . . the Federal Communications Commission, in its administration of the provisions of the Communications Act of 1934, as amended, and as supplemented by this Act, shall . . . (7) grant appropriate authorizations for the construction and operation of each satellite terminal station, either to the corporation or to one or more authorized carriers or to the

> corporation and one or more such carriers jointly, as will best serve the public interest, convenience, and necessity. In determining the public interest, convenience, and necessity the Commission shall authorize the construction and operation of such stations by communications common carriers or the corporation, without preference to either; . . . (Sec. 201 (c) (7).

At first, the Commission held that during an initial period Comsat should be the exclusive owner and operator of the three U.S. ground stations then contemplated. Subsequently, when it was determined that additional ground stations would be needed, the Commission modified its earlier position. It decided that each ground station should be owned 50 percent by Comsat and 50 percent by those carriers which would be served by the station. The ownership ratio among the carriers is determined by the ratio of anticipated traffic handled by such carriers. The stations are operated by Comsat under the policy direction of a committee of the owners.

There was also controversy over whether Comsat should be permitted to deal directly with the government and the large users, or whether it should be a carriers' carrier. The legislation delegated to the Commission authority to decide the circumstances under which Comsat might lease channels to the carriers or might deal directly with particular classes of customers, including U.S. government agencies:

> [Comsat] "is authorized to . . . furnish, for hire, channels of communication to United States communications common carriers and to other authorized entities, foreign or domestic; . . . contract with authorized users, including the United States Government for the services of the communications satellite system; . . ." (Sec. 305 (a) and (b)).

The Commission decided that except in those situations where the national interest dictated otherwise, Comsat would be limited to the role of a carriers' carrier. In compliance with the Commission's decision, a lease agreement between the Department of Defense and Comsat involving 30 circuits from Hawaii to Far Eastern points was assigned to the regular carriers which thereupon charged the Department a composite cable-satellite rate less than the total of what the Department would have paid to Comsat for the satellite circuits and to the carriers for the cable circuits.[21]

As in the case of high frequency radio, the government was anxious to promote the fastest possible development of the new communications technology. The legislation called for competition between satellites and other modes of communications and delegated to the Commission the decision of how such competition should be carried out. The Commission addressed itself to that question in a case involving an application filed by several overseas carriers to lay a new cable between the United States and Southern Europe. This application was objected to strenuously by Comsat. The Commission overruled Comsat's objections and decided to authorize the carriers to construct TAT-5 on condition that they would reduce by at least 25 percent their telephone and telegraph rates and would fill satellite and cable facilities at the same proportionate rate so that 100 percent utilization of these facilities would be reached at approximately the same time.[22]

In summary the following structure of the U.S. communications industry has evolved with regard to overseas communications:

(1) All major U.S. overseas telephone and telegraph carriers use satellite circuits leased from Comsat;
(2) U.S. overseas telephone and telegraph carriers own varying amounts of Comsat stock;
(3) Except under unusual circumstances, Comsat may not lease circuits to ultimate users but is limited to leasing circuits to the overseas carriers;
(4) Major U.S. overseas telephone and telegraph carriers jointly own and use undersea telephone cables;

(5) AT&T, the largest U.S. domestic telephone carrier, is the principal provider of U.S. overseas telephone services;
(6) Western Union, the only domestic telegraph carrier, may not provide overseas telegraph services;
(7) Only U.S. overseas telegraph carriers may offer overseas alternate voice/data services.

Over the years the problems inherent in this structure have been studied and re-studied by the companies involved, by the government agencies immediately concerned (particularly FCC, State, Defense, and Director of Telecommunications Management), and by the Senate Commerce Committee. Various proposals have been developed aimed at changing the overseas communications structure of the industry by bringing about consolidations among the companies involved. However, as conditions have changed, these proposals have been modified.

Earlier proposals have been limited to possible mergers of two or more of the U.S. overseas telegraph carriers. Later proposals contemplated the inclusion of Western Union in such a consolidation. Such a proposal would lead to the creation of a single U.S. telegraph carrier for domestic and overseas telegraph communications. Alternatively it has been suggested to include all operations devoted to rendering overseas services (i.e., AT&T's telephone cable and Comsat's satellite operations). This would lead to the creation of a single chosen instrument for all types of overseas communications. Still another proposal has envisaged competition in the overseas communications field between two companies: (1) a new overseas carrier which would be owned by existing U.S. overseas record carriers, and (2) AT&T. Both the new carrier and AT&T would be permitted to offer all types of communication services. AT&T and the independent telephone companies would be required to interconnect their domestic facilities with those of the new international carrier.

All these proposals have been studied by a Task Force on

Communications Policy appointed by President Johnson on August 14, 1967. The report prepared by the Task Force was submitted to President Johnson shortly before his term of office expired. It recommends the establishment of a single chosen instrument. President Nixon has ordered an evaluation of the Task Force report. This evaluation had not been completed by the time this study went to the printer.

Federal Intercommunications Systems

The largest "private system" consists of the several intercommunications systems used by a number of Federal departments and agencies. The annual total cost of these systems is estimated to be $4 billion. The facilities used for ordinary communication services are leased from commercial carriers. Government-owned communication facilities are limited to providing communication services for certain specialized purposes of the armed services and certain other agencies.

Substantial efforts have been made since 1963 to coordinate the several department and agency systems of the Federal government. During that year President Kennedy, in the wake of the Cuban missile crisis when international communications available to the President were found to be inadequate, directed the establishment of a National Communications System to provide better communications support to critical functions of the government. The Director of Telecommunications Management was asked to provide policy direction for the System, and the Secretary of Defense was designated as Executive Agent of the System. The principal agency systems included within the National Communications System are those of the Departments of Defense, Transportation, State, Interior, and Commerce, the National Aeronautics and Space Administration, the General Services Administration, the Federal Communications Commission, and the Atomic Energy Commission.

The two largest subsystems of the National Communications System are the Federal Telecommunications System (under the jurisdiction of the General Services Administration) and the Defense Communications System. The latter system is by far the larger of the two. It includes an automatic voice network (AUTOVON) and an automatic digital network (AUTODIN). Both networks are still incomplete. The overseas portion of both AUTOVON and AUTODIN is government owned and operated. AUTOVON's domestic portion is leased from and operated for the government by AT&T while the domestic portion of AUTODIN is leased from and operated by Western Union.

The Federal Telecommunications System consists of a voice-grade switched network leased from AT&T and an Advanced Record System leased from Western Union. The two networks provide communications services for the civilian agencies of the Federal government.

Specialized governmental communication systems can play an important role in the communications system as a whole by stimulating the development of new communication technologies and new commercial applications of such technologies. The synchronous satellite, for example, was pioneered by NASA working with Hughes Aircraft Company. The satellites most recently purchased by NASA — the so-called Applications Technology Satellites (ATS) — are designed to carry out experiments in radio communications as well as experiments in meteorology, navigation, radiation detection, radio propagation, and so forth. Several military satellite systems are also capable of making substantial contributions to the commercial communications art.

CHAPTER III

Mass Communications

In order to understand how the present configuration of the U.S. communications industry with regard to radio and television broadcasting* evolved, we shall have to trace the roles in pre-broadcast days of some of the companies which now figure prominently in that configuration. During the early development of radio technology, key admirals in the U.S. Navy Department saw its great importance for trans-oceanic and maritime communications and thus for the nation's security, trade, and commerce.

Radio Corporation of America: A Chosen Instrument

The admirals were concerned that the rapid development of the new technology be undertaken by a U.S. company controlled wholly by U.S. citizens. The first company formed in the United States for the purpose of providing radio services was a subsidiary of the British Marconi Wireless Telegraph Company, Ltd. The Marconi Company of America was organized in November 1899.

In March 1919 the admirals were apprised of negotiations between the General Electric Company and the British Mar-

---

* Sec. 3 (o) of the Communications Act of 1934 defines broadcasting as "the dissemination of radio communications intended to be received by the public, directly or by the intermediary of relay stations." "Radio communication" is defined in Sec. 3 (b) as "the transmission of writing, signals, signs, pictures and sounds of all kinds."

coni Company involving the sale by GE to Marconi of the exclusive rights to the Alexanderson alternator. The admirals believed that these rights were crucial to the development of the art and sought to preclude transfer of these rights to foreign interests. The officers proposed to negotiate a contract with GE pursuant to which a new company controlled entirely by U.S. citizens would be formed to engage in providing radio communication services.

The proposed contract never was executed. Secretary of the Navy Daniels favored a government monopoly of radio communications and he doubted his power to execute the proposed contract except with the consent of the Congress. Congress never authorized the Secretary to proceed with the proposed contract. In 1919, however, General Electric, Westinghouse, AT&T, and United Fruit,* each of which owned important radio patents, organized the Radio Corporation of America. RCA took over GE's inventions in the field of radio and acquired the assets of the American Marconi Company. It can be assumed that this arrangement met with the approval of the U.S. Navy Department since a representative of the Department took a seat on RCA's Board of Directors without, however, having a vote. The purpose of the creation of the new company was "primarily to give the United States pre-eminence in international radio communications." [1] The company's initial role was limited to rendering radio telegraph and marine radio services and to marketing radio hardware manufactured by GE.* RCA was authorized under its charter, however, to engage in noncom-

* United Fruit had become interested in the development of radio for the purpose of facilitating communications which it relied upon in raising bananas in Central America and shipping them to the United States for marketing.

* RCA was not the only company which was engaged in transoceanic radio communications. Federal Telegraph (the predecessor of Mackay Radio which later became a part of ITT World Communications) operated on the West Coast. It later expanded its activities to the Atlantic and Latin American areas in competition with RCA.

munications as well as in communications activities. The charter authorized the new company not only "to send and receive signals, messages, and communications; and to create, install, and operate a system of communications which may be international," but also "to improve and prosecute the art and business of electric communications"; and, among others, "to . . . manufacture, . . . sell, . . . deal in . . . merchandise of every description"; . . . "to acquire . . . and take over all or any part of the assets and/or all liabilities of any . . . firm"; and "to acquire . . . securities . . . of any other corporation. . . ."

While RCA's original business activity continued to grow, it has steadily diminished over the years in relation to its other business activities. It is a well-known fact that RCA has taken advantage of its broad corporate charter and subsequently has engaged in many activities, including radio and television broadcasting, the manufacture of hardware required for this purpose, and also the manufacture of phonographs, phonograph records, computers, and communication satellites. Most recently RCA became a domestic telephone company in the State of Alaska, and it has entered other fields such as publishing and car rental.

Division of Fields According to Uses of Radio Technology

The advent of broadcasting brought RCA into conflict with GE and Westinghouse as well as with AT&T. The Bell System had sought to make broadcasting an integral part of its business operations. AT&T's concept of the structure of the broadcast segment differed materially from the structure that has developed. It resembled much more the concept of community-supported educational or public broadcasting than the concept of advertiser-supported commercial broadcasting which developed subsequently. AT&T, as was explained to a Bell System Conference in February 1923 by the company's vice president in charge of radio matters, intended to own and operate the stations. Broadcasting associations were to be formed by community business leaders and others concerned with the general public and interested in radio, and the associations would provide all programs. The associations would pay AT&T for expenses incurred in operating the stations plus a reasonable return on AT&T's investments in the stations.[2]

Final settlement agreements which were concluded in 1926 between RCA, AT&T, GE, and Westinghouse disposed of existing patent controversies and called for a division of fields among these companies with regard to the use of radio: GE and Westinghouse were to manufacture broadcast hardware and operate some broadcasting stations; RCA was to operate broadcasting stations and sell broadcast hardware manufactured by GE and Westinghouse; and AT&T was to sell its broadcasting station in New York to RCA with RCA committing itself to utilizing AT&T's lines instead of Western Union's for the purpose of interconnecting the broadcasting stations.

AT&T acquired exclusive licenses under the other three companies' patents for wire and radio telephony as well as wire telegraphy. RCA acquired exclusive licenses under AT&T's patents for radio telegraphy and the manufacture for sale of broadcast equipment. (Heretofore RCA had been licensed to use but not sell such equipment.) Subsequently, RCA achieved independence of GE and Westinghouse both in terms of eliminating stock control by the two companies and in terms of manufacturing its own broadcast hardware to be sold in competition with hardware manufactured by the two electrical companies.

These agreements were attacked in 1930 by the Department of Justice as being in violation of the antitrust laws. A consent decree was entered into in 1932. It left the agreements intact except that it voided those clauses which provided for the exclusivity of patent licenses and which restricted the right of the companies to engage in particular lines of business.

The impact of the 1926 agreements with regard to the structure of the communications industry, however, has continued. AT&T still is not in the broadcasting business but it furnishes wire and radio services to interconnect radio and television broadcasting stations. RCA, GE, and Westinghouse are in the radio and television broadcasting business and also are manufacturing and selling radio and television transmitting and receiving apparatus.

The Structure of the Broadcasting Industry

The intercompany agreements discussed in the preceding section were responsible for splitting our communications system into two segments — a mass communications segment and an intercommunications segment. Until recently, the broadcasting industry has been the sole provider of commercial electronic mass communication services and therefore the sole occupant of the commercial mass communications segment. That role is now challenged by cable television operators and others, as will be discussed in subsequent sections. At this point, however, attention will be focused on the evolution of the present structure of the broadcasting industry. (Educational or public broadcasting is carried on by nonprofit organizations and is included in this study only to the extent that it relies on intercommunications to provide network programs.)

The present structure has been shaped by general public policies adopted by the Congress, implemented by the Federal Communications Commission, and supported strongly by the courts whenever the industry has sought to challenge them. The heart of these policies is a reliance on the initiative of numerous commercial broadcasting entities to produce that variety of radio and television program fare which meets the diverse interests and needs of listeners and viewers in different communities throughout the nation.

In order to achieve this objective, the Commission has allocated radio spectrum space sufficient to make possible the licensing of numerous local commercial radio and television stations. Next, the Commission has sought to distribute those licenses among numerous entities, and at present the Commission seeks to make the multiple ownership rules still more stringent.* Relatively few firms own the maximum permissible number of radio and television stations. Paramount among the firms that own the maximum number are the radio and television networks.

Another Commission rule prohibits ownership of more than one network by any single firm. ABC's status as a third independent radio and television network is traceable to that rule since one of the radio networks which RCA was required to sell became the nucleus of the present ABC radio and television network. The Commission has sought to promote the establishment of additional television networks. So far, however, these efforts have not proved successful.

Cable Television and the Broadcasting Industry

Cable television began in 1949 as an auxiliary to over-the-air television broadcasting. In communities which were too distant from TV stations to permit adequate reception, qualitatively and/or quantitatively, of television signals, enterprising local firms proceeded to construct community antenna television (CATV for short) systems. Such systems consist of

* The maximum number is 7 AM stations, 7 FM stations, and 7 TV stations of which no more than 5 stations may be VHF stations (i.e., the other 2 stations must be UHF stations). The Commission has proposed to stiffen this rule in such a way that a firm which already owns an AM station in a market may not acquire and may not be granted a FM or VHF-TV station in the same market. Corresponding provisions would apply to owners of FM or VHF-TV stations. In others words, the objective of the proposed rules is "one-to-a-customer."

an antenna located preferably at an elevated point near the community to be served, and cables strung onto telephone or electric power poles from the antenna to the homes of subscribers, where they are connected to the home TV receiver. The subscribers pay a connecting charge and a monthly fee.

Most CATV operations are franchised by local communities,* and the maximum fees payable by the subscribers are usually fixed in the franchise. In some states, legislation has been enacted or is pending which treats CATV firms as common carriers and grants regulatory controls over them to state public utility commissions.

Early CATV systems were located in small communities and their cables had 5 channels permitting the dissemination of the same number of TV programs. Present-day systems are capable of transmitting 12 to 20 programs, and an increasing number of systems have been constructed in or near major metropolitan areas. The use of microwave systems in conjunction with CATV systems makes possible the reception and distribution of distant signals. Finally, the availability of as many as 20 or even more channels makes possible the origination of television programs by CATV systems, also referred to as cable casting. This circumstance has led to a change in concept as well as name. CATV now stands for "cable television" instead of "community antenna television."

The FCC's views on the regulation of CATV has changed repeatedly from an original position that it had no authority to regulate CATV to an intermediary position of regulating only those systems using microwave facilities. Subsequently, the Commission determined that it had authority to regulate all CATV systems, and the Supreme Court has upheld the Commission authority.[3]

In regulating CATV, a number of issues and interests are

* CATV facilities constructed by telephone companies and leased to CATV operators have been held to come within the telephone companies' general franchise and therefore do not require a special franchise.

involved. CATV charges subscribers a fee and is unlikely to serve homes situated in rural areas. Therefore, would impecunious persons and persons in rural areas possibly be deprived of free television service if CATV in the long run undermines present commercial television services? If CATV is permitted to offer unrestricted competition with present commercial television services by importing distant signals and by originating programs, would such competition be fair in view of the present copyright situation which permits CATV to import distant signals without assuming any of the copyright expenses incurred by the original broadcaster? [4] Should CATV be permitted to threaten the establishment of additional UHF stations in major markets and thus reduce hoped-for increased competition among broadcast stations in such markets?

There is basic disagreement not only within the FCC but also between the FCC and the Department of Justice over whether cable television should be given relatively free rein to develop as a medium competitive with over-the-air television, or hemmed in by various types of regulation so as to play primarily a supplementary role. The Department has favored the former viewpoint while the majority of the Commission have receded gradually from a position strongly favoring protection of over-the-air television from the threatening inroads of CATV. The Department's point of view received strong support from the Task Force on Communications Policy in a report submitted to President Johnson shortly before the end of his term of office. In making the report public, President Nixon stated that publication of the report did not imply in any way Presidential support for any of the conclusions reached by the Task Force.

In December 1968 the Commission proposed the adoption of rules which would *require* CATV systems to originate television programs on one channel but not on more than one channel as a condition of being permitted to carry broadcast

TV programs. Other channels would be made available by CATV operators for cablecasting purposes on a common carrier basis.[5] The new rules would require CATV systems to obtain retransmission consent from the originating television stations for importations of distant signals. In that manner, the Commission suggests fair competition would be achieved between over-the-air television operators and CATV operators.

In October 1969 the Commission issued liberalized rules with respect to cable casting. The new rules direct CATV systems with more than 3,500 subscribers to begin originating programs by January 1, 1971, and permit these systems to sell commercials during "natural breaks" in film, sports, and other programs. The cable casting of commercials will provide additional revenue for CATV systems and may place them in a position where they can compete with broadcasters for program materials. The Commission also indicated that it would oppose any proposal prohibiting CATV systems from interconnecting on a regional or national basis for any purposes, including the cable casting of entertainment programs. A ban on such interconnections had been proposed at one time in the course of negotiations between CATV and broadcast interests. The Commission's position on interconnections is of particular interest since it comes at a time when the use of satellites for such interconnections is under discussion.

The question of what constitutes CATV regulation which is in the public interest given the various conflicting private interests and competing public policy objectives is at present again before the Congress. Congress attempted earlier to enact CATV legislation but was unsuccessful in its attempts. Entreaties on the part of Congress addressed to broadcasters and CATV operators and their respective trade associations to come to some sort of agreement have failed thus far to produce results acceptable to all the highly diverse private interest groups. Broadcasters are more divided among themselves than are CATV operators, but copyright owners constitute a third, very significant force.

Negotiations conducted in 1964–1965 between the National Association of Broadcasters (NAB) and the National Cable Television Association (NCTA) did not produce any agreement. On May 29, 1969, the terms of a proposed agreement reached by the staffs of the NAB and NCTA were made public. Broadcasters would agree to a liberalization of the present limitations imposed by the FCC on television programs which cable television systems may carry. CATV systems could carry the three network programs plus the programs of no more than three nonaffiliated stations. Broadcasters would also go along with the origination on one channel of programs by cable systems and the carrying of advertising on such channel. The systems, on the other hand, would refrain from interconnecting on a nationwide basis.* CATV systems would pay reasonable copyright fees as determined by the Congress.

The staff agreement was accepted by the NCTA but rejected by the NAB and the copyright owners. Whether an agreement can be reached acceptable to a substantial majority of members of both industry associations, as well as copyright owners and various government agencies, remains to be seen.

The reaching of an agreement has been complicated by the fact that an increasing number of broadcasters have decided to enter the cable television business in addition to their original over-the-air television operations. This circumstance has also raised some new public policy issues. There has been some pressure on the FCC emanating from individual members of Congress to adopt rules which would prohibit outright or under certain circumstances the ownership by broad-

* This aspect of the proposed agreement has been criticized severely by those who expect a nationwide cable television system to compete with the present three television networks.

casters of CATV systems. It has been suggested that the ownership particularly of "co-located" systems be prohibited, i.e., where systems and stations are located in the same communities. Such prohibition, it has been argued, would constitute a needed extension of present Commission rules which prohibit ownership of more than one AM, FM, or TV station in the same community. The Commission has asked for comments from industry on the question of ownership of CATV systems by broadcasters and also by nonelectronic media of mass communications (newspapers, etc.).[6] The question of how many CATV systems may be owned by a single entity nationally, regionally, or in any one state, is likewise a subject of Commission inquiry.*

Cable Television and the Telephone Companies

The rapid expansion of cable television into major metropolitan areas and the increasing use of cables with large channel capacity have engendered conflicts not only with broadcasters but also with telephone companies. During the early stages of CATV, telephone companies rented space on their telephone poles to CATV operators to attach their cables. As it became apparent, however, that these cables might be used in the future for two-way, switched broadband services in addition to the present one-way television services,[7]

* The largest system resulting from a merger, announced on August 8, 1969, of H&B American Corporation, Beverly Hills, California, into Teleprompter Corporation, New York, will serve 357,000 CATV subscribers, or almost 10 percent of the estimated total of subscribers in the United States. Hughes Aircraft Company, one of the principal builders of communication satellites, is reported to be Teleprompter's principal stockholder with a 17 percent interest. It is interesting to note that Hughes made an unsuccessful bid for the ABC network and subsequently acquired the Sports Network and Filmation Associates, a major producer of animated films. Mr. Irving B. Kahn, President and Chairman of Teleprompter, is one of the principal protagonists of program originations by CATV systems and of interconnections between such systems to form CATV networks. (Satellites, of course, might be used to interconnect CATV systems.)

some of the telephone companies began to reconsider their positions.

Bell System companies have proceeded increasingly to string their own broadband cables and to lease such cables to CATV operators for rendering traditional community antenna television services. Bell companies may not themselves provide such services. Such services have been held by the FCC not to constitute communication common carrier services, and Bell companies, as will be recalled, are limited under the provisions of a consent decree to providing communication common carrier services.[8]

Leasing cables from Bell companies for CATV operations has certain advantages. Less capital is required; Bell companies install and maintain the cables; but, most of all, it has been held by some courts that CATV operators leasing from Bell companies do not have to obtain local franchises since the Bell companies' general franchises include the leasing and use of broadband cables. On the other hand, because of limitations in their leases, CATV operators who do not own their own cables, and who are not independently franchised, do not have the same freedom of action regarding new broadband services which they may want to offer and therefore they may not reap the same financial benefits as other CATV operators when disposing of their businesses.

The independent telephone companies are not subject to the restraints imposed on the Bell companies by the consent decree. They may, therefore, not only own and lease to others the broadband cables required for cable television services, but may themselves engage in providing such services. At least two of the independents have proceeded to provide such services through CATV subsidiaries. The CATV systems owned directly or through subsidiaries by United Utilities and Continental Telephone are located almost entirely in communities which are served by telephone operating subsidiaries owned by these two companies. The companies have stated that they anticipate providing broadband services other than the transmission of television signals over these systems.

On March 21, 1968, United filed with the Federal Communications Commission and with 11 state commissions tariffs for "wide spectrum services." These services would include CATV; computer data transmission; picture phone; facsimile; automatic meter reading services for local public utilities; and fire, police, and medical alarms on private premises or public thoroughfares. United has stated that it was researching the feasibility of utilizing broadband transmission for regular telephone service so that all communication and information services, including telephone, then could enter business and residential premises by means of a single coaxial cable the size of a lead pencil, normally buried underground.

The FCC has held that cable systems may not be constructed by telephone companies without FCC authorizations pursuant to Sec. 214 of the Communications Act. The Department of Justice has urged that the FCC require all telephone companies to offer pole space to all applicants on equal and nondiscriminatory terms, that telephone companies be prohibited from restricting the uses CATV operators may make of leased cables, and that telephone companies may not offer CATV service in territories served by such companies. In addition, the Department has raised the question whether, in order to "encourage healthy competitive development of the vast potential of broadband coaxial cable," telephone companies should not be precluded altogether from constructing such cable systems until and unless the Commission is satisfied that limitations imposed on telephone companies are adequate to deal with the competitive problems of CATV systems in relation to telephone companies.[9] In view of the present unfavorable regulatory environment, at least one of the independent telephone companies, United Utilities, has revised its earlier plans and is planning to sell its CATV systems serving 70,000 customers in 60 communities.[10]

The future relationships between cable television companies, on the one hand, and broadcasting, telephone, and computer companies, on the other, are likely to present numerous structural problems. The Commission has requested comments on how these several industries see these problems and what solutions they propose. Since the questions which the Commission has propounded constitute a fair sample of the kinds of structural questions which are likely to arise, they are presented in Appendix A.

Pay Television and the Broadcasting Industry

Our present commercial radio and television system depends on advertisers for financial support. An alternative to advertiser-supported television (euphemistically referred to as "free TV") is viewer-supported commercial television (commonly referred to as pay-TV or subscription television). The term pay-TV, as commonly used, implies payment of per-program charges by subscribers who may be charged in addition a small basic monthly fee and an initial connection charge. Several forms of pay-TV are technically feasible.

Over-the-air pay-TV involves the broadcasting of "scrambled" signals which are "unscrambled" by means of electronic devices attached to the home receivers. Scrambling prevents the viewing of pay-TV programs by persons who are not subscribers of pay-TV services. There are various methods of scrambling and unscrambling signals, determining what programs individual subscribers select, and collecting per-program charges for such programs. Zenith Radio Corporation, one of the developers of a patented over-the-air pay-TV system called "Phonevision," has been one of the principal proponents of pay-TV.

Pay-TV may also be operated by cables. Cable pay-TV systems may be built especially for that purpose or cable systems built initially for the purpose of disseminating regular commercial TV programs (the so-called CATV systems) may be used for that purpose.

Operators of advertiser-supported television and motion picture theatre owners have been the leading opponents of pay-TV in any form, whether over-the-air or by cable systems. The FCC has attempted since 1955 to authorize viewer-supported over-the-air commercial television as an alternative method to advertiser-supported commercial over-the-air television. The opponents of pay-TV, however, have found allies in Congress and among the general public who fear that viewers will have to pay for sport and entertainment programs which they are now able to see "free." In response to Congressional requests, the FCC in 1959 limited the number of experiments with over-the-air pay-TV which it would authorize. The application of a Hartford, Connecticut, UHF station to use the Zenith Phonevision system for a pay-TV experiment was the only application for an over-the-air system actually granted by the FCC. The grant was challenged in the courts all the way to the U.S. Supreme Court but was upheld.[11]

At the conclusion in 1968 of the experiment (which proved rather inconclusive because first-run films and major sports events were unavailable), the FCC announced rules to govern the granting of commercial licenses for over-the-air pay-TV on a regular commercial basis. After weighing the arguments for and against over-the-air pay-TV, the Commission rules sought to protect advertiser-supported commercial TV by prohibiting pay-TV operators from showing those sport, film, and series type programs which constitute the main program fare of the former.[12] The legality of the FCC rules has been challenged in the courts. Therefore, thus far no commercial pay-TV licenses have been granted.

The establishment in Los Angeles and San Francisco of cable pay-TV systems failed after a legal prohibition of pay-

TV was adopted in a statewide referendum which subsequently was held unconstitutional by the courts.[13]

Originations of television programs by CATV systems, even though no per-program charges are contemplated, have been opposed because such originations may become an entering wedge leading eventually to actual pay-TV with per-program charges.

PART II

Processes Which Determine the Structure of the Communications Industry

CHAPTER IV

Nature of the Structure-Determining Processes

In Part I we have described the evolution, over a period of one and a quarter centuries, of the present structure of the communications industry. Looking back, we can see that the present structure has resulted not from any master plan but from resolutions of individual conflicts over the distribution among different companies of (1) tasks and (2) opportunities. The tasks are those of engaging in operations over extended periods of time to provide specified communication services, facilities, or hardware. The opportunities are those of securing financial rewards related in some manner to the performance of those tasks.

If all tasks were performed and all opportunities were exploited by a single company — a so-called chosen instrument — we would not have the kinds of structural conflicts with which we are concerned. While the distribution of tasks and opportunities within a single company may produce intra-company conflicts, other companies and governments are not likely to become involved in such conflicts.

If all tasks were performed by the government, conflicts might arise within the government over the distribution of various tasks among different government agencies. In most countries the performance of communication tasks is a government monopoly. However, often mass communication tasks (broadcasting) are entrusted to one government agency while intercommunication tasks (telephone and telegraph) are entrusted to another, usually the postal administration.

One of the most important questions connected with task distribution would then be what opportunities the different government agencies would have to secure funds for performing their respective tasks, i.e., whether they would have to compete with other government agencies for general tax revenues, whether special revenues would be earmarked and set aside for them, or whether they would be able to finance capital expenditures out of the sale of special government bonds.

The occasions for contests among different companies over task and opportunity distribution have varied. Such contests have occurred particularly as new technologies have become available, as new uses of existing technologies have been attempted, as governments have sought to apply new public policies, or as individual companies have pursued new aspirations.

Some future contests over task and opportunity distribution may occasionally still be decided, as some important contests were in earlier years, on the basis of agreements among the contending companies without government participation. However, governments have become involved in most such contests either on appeal by some of the contending companies or because they feel impelled to intervene on their own initiative for the purpose of achieving some public goal.

Government involvement transforms private conflict-resolution processes into public ones. This transformation has the consequence that such processes must provide answers not only to the two questions: — "who *does* what and how" and "who *gets* what and how" [1] — but also to a third question which is exceedingly difficult to answer: — on what grounds does the government answer the first two questions in a particular manner. In other words, government agencies which become involved in conflict-resolution processes must give to the contending companies and the public reasons why they resolve particular conflicts the way they do.

Three elements interact in the public conflict-resolution processes: (1) organizations, (2) reasoned arguments, and (3) procedures. In the discussion which follows we shall describe and analyze separately the characteristics of these elements; but we shall stress all along the interactions that occur among these elements because the interactions among the three elements, and not the three elements separately, make public conflict-resolution processes what they are.

There are several reasons why, in analyzing the working of these processes, we should distinguish between task distribution and opportunity distribution. While the two are related, they are also separable, and in the resolution of some conflicts (with regard to satellites and undersea cables, for example) the task of operating the particular facilities is assigned to a single company while several companies share in the opportunities by jointly owning the facilities. Second, reasoned arguments which are presented and considered in support of, or in opposition to, particular proposals for task distribution are usually supplied by disciplines distinct from those which supply reasoned arguments for the distribution of opportunities. Disciplines concerned with the qualities and the arrangement of materials, and with human and organizational behavior, supply reasoned arguments for task distribution. Disciplines concerned with social, political, and economic relations supply arguments for opportunity distribution. Third, different priorities are assigned by different organizations to the different disciplines and their arguments. Congress and the Department of Justice, for example, are more concerned with the distribution of opportunities while the Department of Defense is more concerned with the distribution of tasks. Fourth, different procedures employed for presenting and considering reasoned arguments affect the respective weights given to reasoned arguments addressed to task distribution or opportunity distribution. Such procedures also affect the range of alternative plans for distributing tasks and opportunities among different companies which may be considered. For example, the selection of more or less formal or more or less adversary procedures in resolving particular structural conflicts is likely to affect the number and the content of the alternative plans which may be presented and considered for distributing tasks and opportunities.

The interaction between organizations, reasoned arguments, and procedures takes place on several levels: *within* organizations (i.e., between suborganizations) and *between* organizations, both private and governmental.

The sum total of the various interactions may be looked upon as a very large, highly complex, and highly changeable open system in operation. The system is open because at any time the borders of the system may be revised as new organizations seek to become parts of the system. The system is highly changeable because at any time the availability of new technologies or new uses of existing technologies may change the relations among the organizations which constitute parts of the system. The system is highly complex because the organizations which are parts of the system are both numerous and highly differentiated.

After this brief discussion of the conceptual tools which we shall use in the analysis of public conflict-resolution processes, we shall now proceed to deal first with the significant characteristics of the organizations which participate in these processes.

Organizations

Communication Companies

The solution of structural problems has become increasingly difficult because the communication companies which interact with one another and with government in determining the structure of the industry have become increasingly numerous and differentiated. By differentiated we mean not only that the companies differ from one another with regard to the tasks which they perform and the opportunities which they exploit; beyond this we mean that as a result of these differences in tasks and opportunities the managers of the companies are likely to have widely differing expectations and apprehensions in planning the future roles of their companies in the face of changing technologies and changing demands. This use of the term "differentiated" constitutes an adaptation of the term as used by Paul R. Lawrence and Jay W. Lorsch in their study of individual firms entitled *Organization and Environment.*[2]

For example, managers of companies which are limited to performing a narrow range of communication tasks and exploiting opportunities connected therewith understandably are defensive about potential reductions of those tasks and opportunities. Vice versa, managers of companies which perform a broad range of both communication and noncommunication tasks and exploit opportunities connected with these are likely to be relatively flexible in their selection.

These differences in expectations and apprehensions on the part of company managers result in different styles of intracompany decisions with regard to the future roles of individual companies within the communications system. They also affect intercompany relations, relations between companies and the financial community, relations between companies and government, and finally relations among various government agencies and various branches of government when they become involved with decisions relating to the structure of the industry.

The Carterfone imbroglio is a case in point. Being restricted to a relatively narrow range of communication tasks and opportunities, though on an exceedingly large scale, AT&T fought hard to retain limitations on foreign attachments of subscriber-owned equipment and interconnections of private communication systems. The managers of the company fought a defensive battle in the belief that the limitations were indispensable, technically to superior performance of the company's tasks, and economically to adequate exploitation of the opportunities connected with those tasks. This attitude brought the company into conflict, not just with Carter Electronics Company, but with numerous actual and potential communication hardware manufacturers, a large segment of the data processing industry, and the operators of private communication systems. Furthermore, it brought AT&T into conflict with various government agencies. Some of them sought to improve the relative position of the government as a user of communications. Others sought to bring about changes of regulatory policies with a view to making the structure of the communications industry more flexible and open.

When, after fighting a losing battle for several years, AT&T was finally compelled to abandon most of the limitations, the company was poorly prepared to visualize what new role it might play in the future within the communications system and to communicate its changed role to its employees, its shareholders, the financial community, and the public at large. As a stopgap, AT&T stressed that its "anywhere, anytime, anything network" was likely to produce greater revenues than before because attachments of a wide range of subscriber-owned equipment and interconnections of numerous private systems would result in more extensive use of the network.

By contrast, the independent telephone companies were not limited to performing a narrow range of communication tasks and related opportunities. Therefore, some of the leading companies have determined to become "total communication companies" by exploiting opportunities connected with a wide variety of communication and related tasks. United Utilities, for example, in its *Annual Report for 1968* stated:

> . . . United expects to expand into activities which promise greater than utility rates of growth and higher rates of return on investment. This portends a change in the character of our company, a change that has meaning for every stockholder and employee, current and potential.

It is interesting to note, as the following table shows, that the financial community considers the opportunities of these independent companies superior to those of AT&T.

Comparative Price-Earnings Ratios of Several Major Telephone Companies
(As of May 31, 1969)

American Telephone and Telegraph Company	14.90
Central Telephone and Utilities Corporation	17.91
Continental Telephone Corporation	21.77
General Telephone and Electronics Company	18.22
United Utilities	22.50

Another case in point of how narrower or wider ranges of tasks and opportunities affect intercompany relations involves AT&T's relations with Comsat as contrasted with ITT's and WUI's relations with that company. ITT's and WUI's managers saw fit to sell substantial portions of Comsat stock originally subscribed by the two companies. The sale produced considerable profits. The managers evidently felt that the companies' financial resources could be put to better use than leaving them invested in Comsat stock. The fact that ITT would lose its representation on the Comsat board did not deter the company from selling the stock. Evidently the managers of the two companies decided that the sale would not hinder the performance of whatever roles they wanted their companies to pursue within the communications system, and would further whatever roles their companies might play outside the system. AT&T's relations with Comsat, on the other hand, insofar as stock ownership and representation on the Comsat board are concerned, have remained unchanged. Circumstances and motivations which might bring about change are likely to be quite different from those of ITT and WUI.

This differentiation among companies which interact in determining the structure of the communications industry becomes even greater if we compare companies which are engaged primarily in providing communication services to others with companies which actually or potentially provide communication services for themselves in support of some noncommunication tasks (secondary communication companies). The steady increase in the number of secondary communication companies and the increased differentiation among such companies have added considerably to the complexity of reaching decisions concerning the structure of the communications industry. The large number of industry organizations representing a variety of such companies in the Carterfone proceeding and the computer inquiry constitutes evidence of the increasing differentiation among the secondary communication companies. This differentiation reflects differences in the quantities and the nature of the communication services which these companies require as well as a difference in financial and personnel resources which these companies have available. The quantities and the uniqueness of the services required by some companies and the great material and human resources available to them constitute incentives for the managers of companies to provide their own services rather than to rely on services provided by the carriers. Other companies merely want to be in a position of using potential self-service as a yardstick in dealing with the carriers with regard to both cost and the kinds of services which are available.

Governmental Organizations

Governmental organizations which interact with one another and with communication companies, in determining the structure of the communications industry, discharge widely differing responsibilities with regard to communication tasks and opportunities. Some of them regulate "in the public interest" the performance of communication tasks and the pursuit of opportunities connected therewith. Others secure communication services, facilities, or hardware for the government's own use, or lay down procurement policy guidelines to be followed by agencies. Still others stimulate the development of new communication technologies or new applications of such technologies.

Governmental organizations on various levels — Federal, state, and local — which belong to various branches of government — legislative, executive, judicial — or which are so-called independent regulatory commissions participate in shaping the structure of the communications industry.

As communications have become increasingly important to the national economy and security and the social and political fabric of the nation, and as new technologies presented new problems, different Federal agencies have been entrusted with various responsibilities in this field. The Mann-Elkins Act of 1910 gave to the Interstate Commerce Commission jurisdiction over interstate and foreign telephone and telegraph services. The Radio Act of 1912 sought to deal with radio spectrum allocation for governmental and private uses but its regulatory scheme proved unable to cope with the great increase in demand for frequencies which occurred in the 1920s. In 1922 the Secretary of Commerce formed the Interdepartment Radio Advisory Committee (IRAC) to allocate frequencies among Federal agencies, and in 1927 Congress established a five-member Federal Radio Commission to allocate frequencies for non-Federal use and to classify, license, and regulate stations to prevent mutual interference.

The Communications Act of 1934 created the Federal Communications Commission, a seven-member independent regulatory commission which was given the regulatory powers previously vested in the Interstate Commerce Commission and the Federal Radio Commission. This transfer of regulatory powers to the FCC left divided regulatory authority with regard to the management of the radio spectrum, with the President having authority to assign frequencies for Federal uses and the FCC having authority regarding all other spectrum uses, including private, state, and local government uses. While the FCC was given specific responsibility for promoting new uses of radio, it was not given similar explicit responsibilities to promote new uses of other modes of transmission nor was it directed specifically to exercise a leadership role in planning and coordinating the development of the communications system as a whole. As in the case of divided authority over spectrum management, the assumption made evidently was that somehow the several pieces would fall into place without deliberate coordination of different communication activities.

When we speak of governmental organizations which *regulate* "in the public interest" we use the term "regulate" in its broadest meaning. We mean to include not only the Federal Communications Commission, but also, among others, the Congress, the President, the Department of Justice, the Department of State, the Department of Defense, and the Director of Telecommunications Management. Congress and the President regulate tasks and opportunities when they pass such legislation as the Communications Satellite Act which provided for the creation of Comsat. The Department of Justice regulates tasks and opportunities when it brings antitrust suits, negotiates consent decrees, or makes formal presentations to Congress or the FCC. The Department of State regulates tasks and opportunities when it introduces foreign policy considerations into the decision-making process, the Department of Defense when it introduces national security considerations, the Department of Health, Education, and Welfare in disbursing funds for educational broadcasting stations and interconnections of such stations, the Director of Telecommunications Management in attempting to coordinate conflicting viewpoints regarding such regulation, the President's Science Advisor in evaluating new technologies, and the Council of Economic Advisers in measuring economic impacts of communication developments.

The governmental organizations which regulate tasks and opportunities are not monolithic in undertaking such regulation. The FCC, for example, consists of seven members who disagree with each other and with different segments of the Commission staff. Such internal division is notable, for example, in the case of the hotly contested distribution of mass communication tasks and opportunities between broadcasters and cable television operators.

Congress, of course, is the very antithesis of a monolithic organization. Numerous committees of both houses of Congress become involved with the various governmental activities which determine the structure of the communications industry: regulating communication tasks and opportunities; securing communication services, facilities, and hardware for the government's own use and for nonprofit educational broadcasting; and stimulating the development of new communication technologies or new applications of such technologies.

The latter activity — stimulating technologies and applications — has been the particular province of the National Aeronautics and Space Administration. NASA has played an important role in the development of synchronous satellites and that development has contributed considerably to the changing structure of the industry. Additionally, NASA has made available Application Technology Satellites for experiments by users with new communication services in order to encourage such users to consider how satellite technology might meet their present and future needs.

Government organizations on the state and local levels are

more likely to become involved as regulators and users in shaping the structure of the industry than as stimulators. State and local governmental organizations, for example, participate along with Federal government organizations in distributing tasks and opportunities connected with cable television and other broadband uses of coaxial cables.

The courts on state and Federal levels become involved because statutes and regulatory decisions dealing with the distribution of tasks and opportunities are brought before them for judicial review of whether particular distributions or barriers to distribution meet constitutional or statutory requirements.

A quasi-governmental organization is the National Association of Railroad and Utility Commissioners which seeks to represent state regulatory commissions vis-à-vis Federal as well as local governmental organizations. With regard to cable television, for example, this association seeks to represent the interests of state regulatory commissions. It has drafted a model state statute which it has asked state legislatures to enact and which would subject cable television to state public utility regulation.

Industrial Associations

In interacting with each other and with governmental organizations, communication companies, hardware suppliers, users of communications, and others feel the need for joining together to advance their common interests. The resulting industrial associations participate to a significant degree in determining the structure of the communications industry. Such participation may take the direction of promoting changes in the structure which exists at a given time or preventing, retarding, or modifying changes which are sought by others.

With increasing differentiation among their members, industrial associations frequently experience difficulties in formulating positions acceptable to all or most of their members. Formulating common positions and securing acceptance of such positions by other industrial associations and governmental organizations require outstanding leadership qualities on the part of the managers of the industrial associations. Toughness and resilience; knowing where, when, and whom to fight; and where, when, and with whom to negotiate — often doing both at the same time — and keeping the members interested and content — these are qualities which often are indispensable for leaders of effective industrial associations.

The industrial associations which seek to play significant parts in determining the structure of the communications industry are numerous and diverse, reflecting the increasing differentiation of companies concerned with communications. In the field of mass communications, there is the National Association of Broadcasters, which includes among its members radio as well as television broadcasters and individual stations as well as group owners and networks. Major independent television stations (i.e., stations not owned and operated by networks), which felt that the NAB did not represent their special interests with sufficient effectiveness, formed a separate organization — the Association of Maximum Service Telecasters — using as the common criterion the circumstance that their members were broadcasting with the maximum power authorized by the FCC. UHF stations are represented by the All-Channel TV Society. Cable television systems are organized in the National Cable Television Association. These four organizations have been in the forefront of both battling and negotiating with regard to determining the future structure of the mass communications segment of the communications industry.

An increasing number of broadcasting companies are owning cable television systems. Some of the managers of these companies have been meeting informally for the purpose of working out a regulatory formula which will permit their companies to have their two cakes while eating them too. These companies constitute a minority within the NAB and the Association of Maximum Service Broadcasters, but their aspirations and apprehensions are factors which are considered when members interact within those two associations and when the two associations interact with the National Cable Television Association and with governmental organizations.

In the intercommunication segment, the U.S. Independent Telephone Association represents the independent telephone companies vis-à-vis the Bell System and governmental organizations. Land mobile systems and private microwave systems are represented respectively by the National Association of Radiotelephone Systems and the Operational Fixed Microwave Facilities. An organization which has chosen a name which suggests rather comprehensive activities — the International Communications Association — has a membership consisting of companies which have plants and offices at various locations and which, therefore, are extensive users of owned or leased point-to-point communication circuits or services (in lieu of circuits).

Many suppliers of communication hardware are members of the Electronics Industries Association. This association includes among its members manufacturing subsidiaries of telephone companies which are to a large extent in-house suppliers of communications hardware, such as Western Electric (AT&T) and Automatic Electric and Lenkurt Electric (GT&E), as well as independent manufacturers which supply hardware to other companies which may be carriers or which operate private communication systems. This differentiation among its members has made it difficult on occasion for the Electronic Industries Association to put forward common positions. The companies which are in-house suppliers of hardware, however, are a distinct minority, and that minority is made up of actual or potential purchasers of hardware produced by the other members. Thus, the members treat each other with a certain degree of circumspection and find it desirable to find accommodations and compromises in dealing with each other and with governmental organizations.

Companies which manufacture business machines, including computers, and which are vitally interested in securing and maintaining liberal attachment and interconnection privileges between their machines and communication facilities, are represented by the Business Equipment Manufacturers Association. Many of the companies which are members of the National Retail Merchants Association operate their private communications systems. The NRMA was the first industry association together with the National Petroleum Association (many of whose members also operate private systems) to intervene in the Carterfone proceedings for purposes of securing liberal interconnection privileges for their members' systems. While such interconnections are of vital interest to member companies which already operate private systems or which may decide to do so at a later time, other member companies may prefer to rely on equipment and services provided by carriers rather than by themselves. Associations like the NRMA, therefore, will seek to protect the diverse interests of both classes of member companies. These interests, however, may come into conflict with each other if at some future date the rates for carrier supplied equipment and services should be raised to make up for loss of any carrier income resulting from liberalized attachment and interconnection privileges.

Other Organizations

In addition to communication companies and government organizations there are other organizations which interact with the first two types of organizations in determining the structure of the communications industry. Among such other organizations are first of all labor unions. Foremost among the unions are the Communications Workers of America with approximately 320,000 members and the Commercial Telegraphers Union with about 28,000 members, both AFL/CIO affiliates.[3] The interests of these unions can be affected greatly by mergers of communication companies and by government rather than private operation of communication facilities. Therefore, the unions are liable to interact with other interested parties when their interests are affected by determinations regarding the structure of the industry.

Professional organizations of lawyers and electric and electronic engineers such as the Federal Communications Bar Association, the Federal Bar Association, the American Bar Association, and the Institute of Electric and Electronic Engineers may occasionally also have an impact, at least indirectly. The legal professional organizations have on occasion concerned themselves with the governmental organizations and procedures which are instrumental in determining the structure of the industry, while the concern of the engineering professional organizations has been concentrated on technical problems, including radio spectrum allocation problems, which are capable of affecting the industry structure in important ways.

Organizations concerned with educational and public broadcasting such as the Corporation for Public Broadcasting, the National Association of Educational Broadcasters, the Joint Council for Educational Telecommunications, and the Ford Foundation have an increasing impact on the structure of the communications industry as special claimants for intercommunications for their broadcasting facilities.

Reasoned Arguments

When we speak of "reasoned arguments" we seek to establish an all-inclusive concept which permits us to group together criteria supplied by various disciplines on the basis of which arguments are advanced why communications functions should be distributed in one way rather than another. Among the disciplines which seek to provide such criteria are law, economics, engineering, accounting, political science, social science, public administration, and business administration. The criteria relate to such diverse aspects as cost and quality of services, reliability of particular facilities, need for coordination between different facilities, company responsibility, innovation, competition, and monopoly. To what extent these criteria are actually determinative with regard to particular distributions of functions, and to what extent they provide rationalizations for distributions sought because of quite different considerations (which frequently are not articulated), will have to be left to another quite different study. Cognizance will be taken here only of the circumstance that these criteria are advanced in the expectation that in some way and to some extent they will be considered and given weight in determining the structure of the communications industry.

Reasoned arguments are divided into two kinds of arguments: special reasoned arguments based on criteria uniquely related to communications and general reasoned arguments based primarily on general politico-economic criteria thought relevant in resolving structural conflicts within the intercommunication segment of the communications industry. The discussion is limited to the intercommunication segment because in the mass communication segment such arguments are addressed primarily to various ways by which radio and television program objectives can be achieved, such as providing adequate opportunities for local self-expression and sufficient program variety to satisfy minority tastes.

Reasoned Arguments Uniquely Related to Communications

There are reasoned arguments which are peculiarly related to communications because they relate to the handling of natural phenomena which are particularly relevant to communications, namely, the radio spectrum. The radio spectrum is a portion of the electromagnetic wave spectrum. The waves have different frequencies.[4] There are portions at either end of the electromagnetic spectrum which are not useful for radio communications: the lowest frequencies are used, for example, for the generating and transmitting of electric power, and the highest frequencies such as X-rays, ultraviolet rays, and infrared rays are used for diagnostic and treatment purposes in medicine and elsewhere.

Within the radio spectrum different frequencies are useful for different types of communications such as point-to-point, broadcasting, and long-distance. As radio technology has progressed we have learned to use higher and higher frequencies but the number of uses has increased even more rapidly. Since different uses of the radio spectrum have the unfortunate tendency of interfering with each other, the spectrum must be considered a scarce natural resource. While we recognize property rights in some other scarce natural resources such as land, we have not recognized property rights in radio frequencies.[5] Instead of being subject to the forces of the market place, the use of radio frequencies is subject to governmental spectrum management processes which consist of *allocating* particular portions of the radio spectrum for specified uses, and then *assigning* particular frequencies within such portions for the exclusive or shared use of specific persons or categories of persons.

Conflicts over allocations of spectrum space for various uses occasionally involve numerous and powerful parties with substantially divergent interests. An outstanding example is the conflict between the television industry, on the one hand, and private companies, states, and local governments over the proposed re-allocation for mobile communications of frequencies now allocated exclusively for UHF television. Alternatively, a sharing of such frequencies by the two classes of services has been suggested.

Mobile communications are used by state and local governments for many purposes including particularly police and fire protection, and by private companies for railroad, truck, bus, and taxi operations, remote control of machinery and industrial processes, and paging services for personnel. When the FCC allocated frequencies for mobile communications in 1949 "with the possible exception of a few unavailing voices, no one then envisioned the birth of so many diverse services and the astonishing growth pattern of the entire group." [6]

The severity of this and several other still unresolved conflicts has prompted a re-examination of the spectrum allocation processes which are presently employed in the United States.[7] Reasoned arguments have been advanced by specialists in various disciplines concerned with spectrum management problems with a view to improving the utilization of the radio spectrum. Engineers seek to achieve more effective spectrum use through application of improved engineering techniques to transmitting and receiving equipment capable of using narrower frequency channels or less crowded channels. They also seek to improve their knowledge of interference characteristics. Economists seek to achieve more efficient spectrum use through such devices as recognizing spectrum rights which may be bought and sold in the market place, levying users' charges, or simulating market performance in making administrative allocations of spectrum space.[8] Lawyers seek to improve legal procedures for determining who may use what portion of the radio spectrum for what purposes.[9] Public administration specialists give their attention to the governmental entities which determine who may use what radio frequencies for what purposes.[10]

While the various specialists are able to advance and debate reasoned arguments relating to their respective specialties, a vast communications gap precludes integrated consideration and resolution of conflicts involving reasoned arguments advanced by different disciplines.[11]

The extent to which the utilization of scarce spectrum space can be improved affects not only radio communications but other modes of communications as well. The scarcer spectrum space becomes, the greater will be the need for saving spectrum space by using modes other than radio communications where possible. Furthermore, the scarcer space becomes, the more the question will come to the forefront whether carrier provision of radio communication services can achieve greater spectrum economies than private operation of radio facilities. Thus, reasoned arguments with regard to spectrum management cut across many of the crucial structural problems in the communications industry. While such arguments are addressed to what are represented as exclusively technical problems, these problems have highly political implications and, therefore, must be dealt with accordingly.

General Reasoned Arguments Applied to Structure of Intercommunication Segment

The general reasoned arguments which are considered in the course of structural conflicts in the intercommunication segment of the communications industry are based on values derived partly from conventional wisdom, partly from economic theory, partly from experience with structural regulation of other politico-economic systems (particularly the transportation system), and partly from technological requirements. It is part of our conventional wisdom (largely politico-economic influenced) not to have faith in the long-range beneficence of monopolies which we consider untrustworthy because they are repositories of concentrated private power. (Engineers often favor the opposite point of view.) On the other hand, we also distrust governmental power and we know from experience that the traditional tools of public utility regulation are crude at best. For example, by making the rate of return which we allow utilities dependent on their rate base, rate regulation is based primarily on historical costs. It rewards high investments instead of providing incentives for effective and adequate levels of performance. Instead of providing incentives for the utilization of more efficient technologies, rate regulation actually tends to discourage such utilization.

Under these circumstances there is a strong desire both to avoid monopolies and not to depend wholly on public utility-type regulation to secure acceptable performance by regulated industries. Efforts are made to achieve acceptable performance of regulated industries by building into the structure of such industries the maximum possible number of competitive features.[12] Such efforts have led to divisions (or attempted divisions) in a number of different ways of functions within the intercommunications segment: (1) performance of identical intercommunication services by more than one company; (2) performance of substitutable intercommunication services by different companies; (3) operation of different modes of transmission by different companies; (4) manufacture of hardware and rendition of services performed by different companies; (5) rendering specialized intercommunication services by specialized carriers or private systems in competition with regular carriers; and (6) performance of communication tasks by carriers' carriers instead of regular carriers.

As shall be demonstrated by means of some examples, by distributing functions for the sake of injecting elements of competition, a risk is run of shifting the focus subtly from acceptable performance of a very large and complex multipurpose system over extended periods of time to the performance of parts of such a system with regard to some single purpose over relatively short time spans. Instead of asking how tasks and opportunities should be distributed between companies in such a way that such companies will have adequate long-range potentials and incentives for contributing to the continuing development of the multipurpose intercommunication system, we seek to ascertain whether at any given point in time a particular class of customers receives the desired quality of some particular service at the lowest possible cost. This was the basic issue on which the FCC majority and minority split in the Microwave Communications case.

In so doing, we overlook the fact that there is no actual market place in which the latter question can be determined, where the successful company furnishing such a service wins the competitive race. Instead, the companies are likely to fight successive battles with each other before government agencies in pursuing particular business opportunities. Such battles tend to overload the agencies and thus tend to produce numerous and prolonged deadlocks.

It would be interesting, for example, to ascertain the amount of time and effort expended by the FCC on resolving conflicts over the ownership of satellite ground stations, and the construction and use of satellites and undersea cables resulting from competition between Comsat and the telephone and telegraph carriers. Similarly, it would be interesting to observe how much time and effort the Commission will have to expend in deciding conflicts over rates charged by specialized carriers such as Microwave Communications and the regular telephone and telegraph carriers for competing services offered by them. In these cases of competition between companies, it will be the Commission and not the market place which will ultimately have to resolve such conflicts.

Under these circumstances, the question is to what extent pluralism of companies is likely to render long-range system performance more acceptable than it might be otherwise, and to what extent such pluralism is likely to make such performance less acceptable because it overburdens the regulatory agencies instead of injecting actual market competition and because it impairs the capacity within the system for reaching decisions with regard to the introduction of new technologies to such a degree that numerous and prolonged stalemates are likely to develop.

It must be recognized frankly that answers to this question can be only approximate since appropriate yardsticks are lacking with which to measure with precision, and predict on the basis of such measurements, what distribution of functions will result in acceptable long-range total system performance, keeping in mind that the system is expected to meet several competing objectives: (1) general public objectives, such as the availability to the general public at any given time of a wide range of rapid, reliable, convenient, and reasonably priced communication services; (2) governmental objectives, such as having available facilities and services adequate to meet objectives relating to economic development, national security, foreign policy, and social advances; (3) special private objectives, such as the availability of opportunities to secure communication facilities or services designed to meet special private business needs on terms acceptable to the users; and (4) the overall objective of keeping the system open at all times for the introduction of technical advances generated from within or outside the system.

When we speak here of "acceptable long-range system performance" we are using a shorthand phrase for "long-range system performance which avoids unacceptable results." Using the double negative repeatedly throughout our discussion would be rather tedious. However, it must be remembered that there is a considerable difference between establishing standards of acceptable performance and establishing standards of unacceptable performance. It is easier by far to do the latter. "Rules are more workable if they are of the 'don't' variety rather than the 'do' variety." [13]

The temptation of shifting the focus from long-range system performance to short-run performance of its parts is furthered by a general tendency to place greater confidence in judgments which appear to be more precise because they are based on factors which are readily quantifiable.[14] Given this tendency, it then becomes a matter of faith that reasonably precise judgments of the short-run performance of several parts, even though when added together they form a total short of the whole, are superior to less precise overall judgments of how well or how poorly the total system is likely to perform over extended periods.

In the following discussion we shall give a few examples of how, instead of trying to make adequate system-wide long-range judgments, the FCC, the Department of Justice, and Congress on several important occasions have focused their sights on short-run performance of various parts of the system.

Competition Between Identical Services — During the early 1950s in the RCA Communications–Mackay case, the FCC held that competition between two companies rendering the same services in two communication markets was necessarily beneficial without scrutinizing closely the impact of such competition on the performance of the communications system as a whole. The case involved an application by Mackay Radio and Telegraph Company for authority to operate overseas radio-telegraph circuits between the United States and Portugal, and between the United States and the Netherlands. The application was opposed by RCA Communications which operated circuits between the United States and those two countries. The Commission decided in favor of Mackay primarily on the grounds that Congress as a matter of national policy traditionally had favored the principle of competition over the principle of regulated natural monopolies. The Commission, therefore, proceeded to affix the stamp "found to be in the public interest" to the conflict resolution which it sought to impose.

The Commission's decision was challenged in the courts by RCA Communications. The U.S. Supreme Court held that the Commission had failed to carry out the Congressional mandate which required it before reaching decisions to weigh carefully in each case all conflicting evidence and all conflicting reasoned arguments relevant to the conflict under consideration.[15] In the words of the court:

> While the Commission recites that competition may have beneficial effects . . . it does so in an abstract, sterile way . . . instead of reaching . . . its independent conclusion, from the impact of the trends and needs of this industry . . . and . . . bringing the deposit of its experience, the disciplined feel of the expert, to bear on applications for licenses in the public interest.
>
> That there is a national policy favoring competition cannot be maintained today without careful qualifications. It is only in a blunt, undiscriminating sense that we speak of competition as an ultimate good . . . it is for us to recognize that encouragement of competition as such has not been considered the single or controlling reliance for safeguarding the public interest. Of course, the fact that there is substantial regulation does not preclude the regulatory agency from drawing on competition for complementary and auxiliary support. Satisfactory accommodation of the peculiarities of individual industries to the demands of public interest necessarily requires in each case a blend of private forces and public intervention.

The subsequent history appears to bear out the Supreme Court's misgivings. Mackay no longer exists as a separate overseas telegraph company. However, there are in existence at present three major overseas telegraph carriers duplicating to some extent services to the same overseas points. In 1964 the FCC accepted the plea of these companies that their continued existence could be assured only by giving them the exclusive right to furnish alternate voice/data services and denying that right to AT&T. Since then, the merger of some or all of these companies has been urged by several government-sponsored studies. One submitted late in 1968 by the Task Force on Communications Policy recommended performance of all overseas communication services by a single carrier. This recommendation was based in part on the conclusion that the benefits derived from intercompany competition were insufficient and that the continued exclusive authorization to furnish alternate voice/data services in order to maintain the companies in operation could not be justified.

The question must be examined whether the Supreme Court's criticism is not applicable also to some of the distributions of different functions within the present communications system effected by other government agencies. Perhaps some of these distributions have also been justified in an "abstract, sterile way" without careful scrutiny whether the competition expected from such distributions actually will bring about improved long-run system performance.

Competition Between Substitutable Services — For example, the provision of different forms of communications, such as record and voice communications, are often viewed as competitive communications functions which should be assigned to different companies in order to secure the benefits of competition for the consuming public. That at least was the principal justification advanced by the Department of Justice early in the history of structural regulation of the industry when it insisted upon AT&T's relinquishing control of Western Union.

In dividing communication functions in this particular manner, the Department sought to prevent AT&T from becoming the sole provider of all the communication services then known. While the Department achieved a negative antitrust objective — namely, containing the expansionist objectives of AT&T — the more crucial question is whether it also achieved the vital positive objective of so distributing tasks and opportunities among major companies that each of them has the potential of contributing substantially to the development of the intercommunications system.

Applying that positive test, we are compelled to conclude that the distribution which resulted in maintaining Western Union as a separate business enterprise, engaged primarily in providing public message telegraph services, has not enabled that company to contribute significantly to the development of the intercommunications system. Shortsighted government policies, such as FCC's failure to support appropriately at an early stage Western Union's plans for a nationwide microwave system, are to blame in part for this result. Basically, however, Western Union's principal assigned task of providing public message telegraph services has been inadequate to make the company viable. While such services provide users with a written record of the transmitted information, they do not give complete assurance to the users that their messages will be received nor do they provide them with convenient instantaneous two-way communication between persons or machines. The subsequent development of switched teletype services, which provides such communication services, and the concentration on the furnishing of such services exclusively in Western Union, may finally result in that company's being cast in a more viable role within the communications industry. This may come to pass especially if the furnishing of these services can be combined with furnishing new types of computer-communication services. Such combined services, however, present new problems insofar as competition between Western Union and computer service companies is concerned. These problems in 1969 were receiving the attention of the FCC and the Department of Justice.

Competition between substitutable services was also involved in the Microwave Communications case where this company proposed to offer two kc voice channels with sharing privileges by several subscribers while the telephone carriers offered four kc voice channels. In that case the FCC majority held that the two kc channels would provide an acceptable though lower quality service at lower cost, and that potential subscribers should be given a choice between the two types of service. The FCC minority, and particularly Chairman Hyde in his dissenting statement, pointed out that the "talismanic reliance" upon competition favored by the majority was likely "to cost the average American ratepayer money to the immediate benefit of a few with special interests." [16]

Competition Between Different Modes of Transmission — If a similar dual test — negative and positive — is applied to the distribution of tasks and opportunities within the communications industry in connection with satellite communications, it may be concluded that while that distribution to some extent meets the first test, it is likely to fail in meeting the second, infinitely more difficult test. The establishment of Comsat was designed to preclude, on the one hand, government operation of satellites and, on the other hand, the expansion of AT&T's activities, on an exclusive basis, into satellite communications. Comsat's establishment was also designed to achieve the further antitrust objective of assuring access by nonintegrated hardware manufacturers (primarily aerospace manufacturers), and to preclude limiting the market to integrated carriers like AT&T, GT&E, RCA, and ITT. However, when it comes to performing communication tasks, Comsat performs only one element — the long-haul element — in connection with the furnishing of general instantaneous two-way communications. If Comsat were owned, as was originally contemplated by the FCC, exclusively by the carriers, Comsat's limited role of being a carriers' carrier would not present particular problems in relation to the carriers. Being owned, however, originally

50%, and presently in excess of 60%, by the general public, Comsat's role is a contradictory one. If the company continues to be limited to being a carriers' carrier and the chosen instrument for U.S. participation in Intelsat, its role as a profit-making business enterprise is not likely to be a viable one over extended periods of time unless the capitalization is reduced materially. Therefore, of necessity, Comsat has become a company in search of adequate tasks and opportunities, because the opportunities which are connected with its originally assigned tasks do not seem to the company adequate in the long run to make it a viable commercial enterprise.

Some of the notions with regard to the "competitiveness" of modes of transmission and different services (record, voice, data) have been borrowed from the field of transportation. Reference is made not infrequently to "competition" among modes of transportation and among different transportation services as evidence that similar "competition" should exist among modes of transmission and different communication services. It should not be forgotten, however, that the distribution of tasks and opportunities within the transportation industry on the basis of "competition" among modes and services has necessitated government intervention through the creation of the United States Department of Transportation for the purpose of providing a greater degree of coordination in providing the ultimate transportation services which are required by the general public and by governments to move goods and persons from door to door. There is considerable support for the point of view that the transportation system fails to perform in an acceptable manner because adequate coordination between different modes and between complementary functions has been lacking, and that therefore the system fails to provide the desired ultimate complete transportation services.[17] Under these circumstances, the practice of urging the distribution of tasks and opportunities between "competing" companies in the communications system on the grounds that something similar has been done in the transportation system should arouse more than a little suspicion.

The crucial problem, then, in communications even more than in transportation is to determine the respects, the extent, and the forms in which "competition" and "coordination" are required to assure that the communications system will perform in an acceptable manner over extended periods of time. In making these decisions, we must focus at any given time on all of the ultimate complete services — old and new — which the communications system is expected to provide to the general public, governments, and special private users. In distributing tasks and opportunities, therefore, we must have regard for the sum total of ultimate services involved, and we may not limit our focus to particular parts of the system because *conscious coordination* of the different parts is required to provide all the ultimate services.

This point has been stressed in the Dissenting Statement, dated December 10, 1968, which James D. O'Connell, Director of Telecommunications Management, appended to the Final Report of the Task Force on Communications Policy submitted to President Johnson December 7, 1968:

> There are two general themes which run through most of the report. The first is the need for more competition; the second, the need for greater innovation. I have no disagreement whatever with these objectives, but I disagree with the philosophy that these are ends in themselves as they are applied to the telecommunications common carrier service. Fundamental goals are improvement of service and reductions in cost, and these goals are being progressively achieved. There is no question that competition, properly employed in an appropriate market place, can and does produce economic benefits, stimulates innovation, and minimizes the need for Government intervention or regulation. But the question before us in the furnishing of telecommunication services is

the nature of the market place, the record of the results produced, and the promises and problems of the future. In our judgment, the unique and peculiar nature of telecommunications services makes it of paramount importance that the record, the promises, and the prospects of competitive benefits be critically examined.

It is a faulty conclusion, in our opinion, that we can automatically move more rapidly toward improved telecommunications services either through the expedient of creating more competition between communications carriers or by introducing greater innovation in communications hardware in this highly complex national telecommunications structure. It is perhaps worthy of note that while Chapter Two of the Report recommends the advantages of consolidation of competing long haul facilities to rectify numerous difficulties, Chapter Six proposes the creation of new competing long haul facilities.

There is no lack of appreciation or enthusiasm for competition when it can, and does, produce desirable results in the field of innovation, improvement of service, and cost reduction. However, there is a long history of the adverse effects upon the public interest during the years of intense competition in the telephone industry. It is important to consider the long record of judicial and administrative decisions, which suggests that our overall national philosophy of the unfailingly beneficial results of competition cannot be applied across the board to the telecommunications service sector of the national economy. It is one conclusion of this dissent that all proposals for increasing or decreasing competition in this industry should be examined much more closely in the light of past history and in greater detail as to future implications and effects than has been possible within the time frame of the Task Force study, and that the experience and views of all elements of the industry involved should be given more consideration.

Competition Between Hardware Manufacturers — The issue of "competition" versus "coordination" is also involved prominently in alternative schemes which have been advocated for the distribution of tasks and opportunities with respect to (1) providing communication services and (2) manufacturing communication hardware. Contentions have been advanced on an all or nothing basis, particularly by some carriers, that the complete integration of these tasks and opportunities is "right per se" and that any separation would spell disaster for the proper functioning of the switched telephone network. Opposing contentions have been advanced from time to time, particularly by the Department of Justice, that complete separation of these tasks and opportunities would greatly increase competition among hardware manufacturers and presumably, therefore, improve the performance of the system as a whole.

Considerable disagreement exists among antitrust experts in what situations the antitrust laws should be applied in any industry to vertical integration and product extension. Experts who insist on limiting the antitrust laws to achieving economic objectives exclusively advocate noninterference with companies' freedom to integrate tasks which they consider complementary. Experts who look upon the antitrust laws as being designed to achieve multidimensional (i.e., social as well as economic) objectives, are inclined to object to vertical integration and product extension both on social grounds (concentration of private power) and on economic grounds (increased barriers to business entry).[18]

In the case of the communications industry, the disagreements engendered by these different approaches are compounded further by the need for great exactitude in coordinating manufacturing and service functions performed by different companies in order to provide the desired ultimate services. Whether the required coordination should be provided internally by a single management or externally by prescribed procurement procedures, official technical standards, or other coordinating devices is a question on which companies, agencies, and academicians are often in fundamental disagreement.

The distribution of hardware and service tasks, however, instead of presenting questions of principle which may be answered categorically one way or another, have to be determined on the basis of numerous variables which have to be balanced, and the balance which may be struck in the public interest is likely to differ under different circumstances. For example, the attachment of customer-owned hardware in the case of residential telephone services poses questions of equipment maintenance quite different from those presented in the case of private communication systems, operated by larger business concerns, which are capable of providing their own maintenance services or else retaining qualified firms for this purpose.

In the case of the manufacture of communications hardware by nonintegrated manufacturers, the ability of carriers to write adequate specifications, to set proper standards of performance, and to know what prices to pay are some of the factors which should be considered in distributing manufacturing and service tasks between carriers and manufacturers. The consideration of those factors is likely to lead to the conclusion that in order to enable the carriers to do these things they should have some in-house manufacturing capabilities. How extensive such capabilities should be will have to depend on such factors as economies of scale which may be achieved in manufacturing particular products, and the number of alternative sources of supply which are economically feasible for specific products.

Those who argue for separating, on principle, manufacturing tasks from service tasks usually point to other industries as "evidence." One of these industries is the air transport segment of the transportation industry. The separation of airline operations from aircraft manufacture was made mandatory by law[19] because there was specific evidence that vertical integration had led, in the case of a specific airline, to the use of aircraft manufactured in-house only, and that the particular aircraft was unsatisfactory. This instance may be considered evidence by some that all integrated manufacturing and service operations are suspect and should be outlawed as a matter of principle. Before reaching this conclusion, however, it may be desirable to look at some countervailing "evidence." Richard E. Caves in his study of *Air Transport and Its Regulators*[20] has pointed out that because of the understandable desire of competing aircraft manufacturers to sell new and improved passenger planes to the airlines, and because of the willingness of the airlines to acquire such planes, the operation of new planes has become one of the principal ways in which airlines compete with each other. According to Caves, close cooperation exists, with the consent of the CAB, between airlines and aircraft manufacturers with regard to the development of new passenger planes. Caves comments that the acquisition of new equipment by the airlines has been favored by CAB policies which make possible rapid amortization of such equipment. Caves questions, however, whether less rapid and less extensive replacement of planes, coupled with charging of different fares on new and old but still usable equipment, might have been a preferable public policy. In the spring of 1969 the CAB turned down a request by some airlines for higher fares on the grounds that reduced purchases of new equipment would enable the airlines to maintain the present level of fares.[21]

Under these circumstances, there may be some question whether the distribution of hardware and service tasks in the air transport industry does not have some possibly undesirable side effects: high levels of passenger fares made necessary by rapid equipment turnover stimulated by strong competition among aircraft manufacturers. It would seem that such possible side effects must be taken into consideration in determining the proper allocation of hardware and service tasks within the communications system. In other words, it cannot be assumed that the long-range performance of that system as a whole will be improved automatically if more competition is introduced into some part of that system.

In any event, competition with regard to hardware, if introduced, is likely to have little resemblance to an open market. Since coordination between hardware and service is required, close contractual relations over extended periods similar to those which exist between airlines and aircraft manufacturers are likely to result. The establishment of suitable technical performance standards may promote the development of a broader and more open hardware market. In spite of such standards, however, purchasers of hardware are likely to insist that the suppliers have "credibility"; i.e., they must demonstrate their capabilities with regard to hardware design, their reliability in meeting standards, specifications, and delivery schedules, and their continuing availability as sources of supply. Demonstrations of such credibility by new firms without records of past experience or by established firms entering fields new to them are not easily furnished. Purchasers, therefore, are likely to prefer dealing with their regular suppliers.[22] The crucial question, therefore, is to what extent and in what way long-range antitrust objectives of keeping the system open to technical advances are achievable by separating manufacturing and service functions while at the same time preserving management responsibility for rendering reliable services at reasonable rates.

On the other hand, there can be no doubt that the distribution of hardware and service tasks and opportunities among different companies constitutes one of the crucial problems which has to be faced by industry and government. Present hardware requirements of the communications system are huge, and with the expansion and diversification of the system they are growing. There are relatively few companies which seek to provide communication services. However, the number of companies which seek to provide communications hardware is large and growing. Companies which have not been hardware suppliers in the past demand a piece of the market, and companies which have been suppliers seek a larger piece. The Commission's Carterfone decision which liberalized foreign attachment and interconnection privileges has opened up new opportunities for suppliers to sell customer-owned attachments and hardware required for private communication systems. On the other hand, there is no comparable governmental policy decision which has the effect of opening up increased opportunities for hardware suppliers to sell to integrated carriers. These constitute the bulk of the hardware market and at present they are to varying degrees self-suppliers of the required hardware. According to information furnished by the telephone companies, the rate of in-house supply of hardware ranges from 85 percent in the case of AT&T down to 15 percent in the case of some of the larger independent telephone companies. If independent hardware suppliers are largely unsuccessful in gaining better access to the main market, they are likely to press, with the support of the Department of Justice, for the expansion of markets constituted by private systems, specialized carriers, and carriers' carriers to compensate for their inability to enter the chief market.

The support of the Department for such efforts has been in evidence in several FCC proceedings involving such peripheral markets. Such support may be due in part to the uncomfortable position in which the Department finds itself with regard to securing access for independent manufacturers to the chief market. As will be recalled, the Department originally sought to open up this market through the outright divestment of Western Electric by AT&T but then undertook to negotiate a consent decree limiting AT&T to communication common carrier activities but permitting Western Electric to continue to supply AT&T's hardware needs in connection with such services.

In view of the provisions of the consent decree, the Department evidently feels that access to new markets for independent hardware manufacturers should be secured by creating peripheral markets to compensate for the closed main market. The most recent example of the Department's attitude in this respect may be found in its brief filed with the FCC in the proceeding dealing with CATV operations by telephone companies. The brief stresses the role of independent CATV operators as potential customers for nonaffiliated hardware suppliers and therefore seeks to limit the entry of telephone companies into CATV operations. This raises the question whether excessive expansions of private systems, specialized carriers, and carriers' carriers are likely to lead to distortions of the system as a whole with the attendant risk that the system is less likely to function in an acceptable manner in the longer run. This question comes up particularly as the Department urges the FCC to give serious consideration to precluding telephone companies from offering new services such as broadband services. The pursuit of policies precluding telephone companies from utilizing new technologies to provide new services can only lead to giving such companies a "railroad-mentality" with the resulting adverse consequences for the long-run performance of the system as a whole. In view of these dangers for the telephone companies as well as the system, it would be preferable for the companies and the government to adopt policies which will tend to open up in a satisfactory manner the chief market for communications hardware and which will leave the companies free to utilize new technologies to provide new services, including particularly two-way switched broadband services.

The problem which has to be faced here by the integrated carriers and by government regulators is not unlike the problem faced by Intelsat and Comsat in connection with international satellite communications. In that case, European, Japanese, and other foreign hardware manufacturers are seeking an increasing share of the satellite hardware market, a very large percentage of which has been supplied by U.S. manufacturers. To the extent that the foreign manufacturers are unable to achieve satisfaction of their demands, they will seek to promote regional satellite systems to which they expect to have better market access than to the international system. The maintenance and expansion of the worldwide international system, therefore, depends on arriving at a negotiated compromise formula, which on the one hand is designed to meet at least some of the demands of non-U.S. hardware manufacturers, but which, on the other hand, does not jeopardize the expeditious development and economic functioning of the worldwide system.

Competition Among Common Carriers, Specialized Carriers, and Private Systems — One of the criteria for determining what constitutes "acceptable" performance is how well the system accomplishes the balancing of two goals: (1) providing services sought by the general public and (2) providing services demanded by special private business groups. One school of belief advocates that the provision of bulk and specialized business services by specialized carriers, or by means of facilities owned and operated by the providers of such services, will *per se* improve the performance of the system as a whole because competition within the system will be increased. The opposing school contends that any such provision of services will lead to "cream-skimming" and compartmentalizing the system and, therefore, will result *per se* in poorer performance of the system. Neither school is prepared to substantiate its contentions regarding the likely consequences for the performance of the system as a whole, but both schools are equally insistent upon proclaiming their beliefs.

Actually, the distribution on a rational basis of tasks and opportunities between integrated carriers, specialized carriers, and self-providers of services turns on such questions as the degree to which the averaging of rates and the standardization of services and equipment are desirable. Answers to these questions depend on the particular services involved and the impact on the system as a whole of providing such services one way rather than another. Under these circumstances, a *rational* distribution of tasks and opportunities means examining the likely technical, economic, social, and other consequences of alternative distributions for the performance of the system as a whole, and reaching decisions accordingly. This does not mean that a particular distribution decided upon will necessarily be the "best" one. We shall never know whether or not it is. It only means that hopefully we will avoid selecting from among suggested alternative solutions those extreme ones which on the basis of our analysis are likely to lead to undesirable consequences with regard to the future performance of the system as a whole.

Competition Between Geographically Limited Monopolies — The retention of geographically circumscribed monopolies in the form of independent telephone companies separate and apart from the Bell system was one of the policy objectives of the government's earliest structural regulatory efforts which culminated in AT&T's Kingsbury Commitment. A geographically circumscribed monopoly was also recommended in December 1968 by President Johnson's Task Force on Communications Policy with regard to all U.S. overseas communications.

Two of the great advantages of distributing tasks and opportunities solely on a geographical basis flow from the fact that (1) the companies involved can offer all conceivable communication services and (2) they are more ready to recognize and accept in their own best interest and in the public interest the need for coordinating their respective operations, particularly by establishing mutually acceptable standards. The recognition and acceptance of such need are required equally in the case of other distributions of tasks and opportunities. Considerable changes in attitude on the part of the principal companies involved will be necessary, however, before such need will be similarly recognized and met in the case of other distributions. Under these circumstances, divisions of functions on a geographical basis appear to have substantial long-run advantages not present in the case of other distributions, provided that the companies are large enough with adequate resources to play significant roles in the longer run in developing the communications system.

Government-Sponsored Companies — While this completes our review of various schemes for distributing tasks and opportunities within the communications system, a few observations are appropriate on the use of the device of government-sponsored companies for the purpose of discharging new tasks and exploiting opportunities connected therewith. This device was employed when the government sponsored the creation of RCA, and it was used again in the creation of Comsat.

Conventional wisdom supports the general proposition that new organizations are frequently needed to apply and exploit new technologies because existing ones are often reluctant to do so. This proposition is based on observations of actual incidents where existing organizations failed to take advantage of new technologies in spite of the availability of opportunities to do so. Western Union's unwillingness to purchase Alexander Graham Bell's telephone patents, and the refusal of companies engaged in the manufacture of photographic and duplicating equipment to acquire patents relating to what later became known as "xerography" are cited as outstanding examples.

A related proposition is based on the observation that new technologies are often developed by companies involved primarily in activities lying outside the field in which such technologies are later applied and exploited. The development of jet engines by airframe manufacturers rather than by manufacturers of aircraft engines[23] and the development of diesel engines by automobile manufacturers rather than manufacturers of steam engines are examples in point. The development of synchronous communication satellites by a space equipment manufacturer rather than by a communication equipment manufacturer is also suggested as an example.

Unequivocal judgments cannot be made that sponsorship by government of new companies to discharge new communication tasks constitutes either a suitable or an unsuitable organizational device. Such judgments must depend on many factors, including particularly the time horizons which we apply in passing judgment, and whether we focus on the performance of the companies themselves or on the impact which their existence has on the performance of the system as a whole.

Taking a longer-range view, the scope of the charters which such companies are granted is highly important. RCA's charter was broad in that it included noncommunication as well as communication activities while Comsat's charter is limited to satellite communication activities. Using political terms, RCA's broad charter afforded that company ample "Lebensraum" and enabled it gradually to throw off the shackles imposed on it by other companies which sought to promote their own interests by limiting RCA's aspirations. Comsat's narrow charter, on the other hand, established that company as the organizational equivalent of a "buffer state." It was created largely to provide a buffer zone between the communication carriers and the aerospace industry. The short-run advantage of Comsat's existence has included (1) a satellite-oriented management which has successfully applied synchronous satellite technology to provide worldwide international communications, and (2) the stimulation of great interest in applying that technology to provide other services.

However, a price will have to be paid in the longer run for that advantage. Continuing conflicts will have to be resolved over who shall be permitted to own ground stations, and the extent to which undersea cables and satellites may be used to provide communications services. Even more important, however, the creation of a separate company limited to exploiting opportunities connected with a single mode of transmission tends to complicate, and therefore delay substantially, the assignment of tasks and opportunities connected with the use of satellites for domestic communication purposes.

On the other side of the ledger, the most significant beneficial long-range impact of Comsat is likely to be an indirect one: there is apt to be in the future a greater willingness to experiment with new organizational devices for distributing tasks and opportunities, and individual companies may prove sufficiently daring to utilize such devices to furnish jointly new communication services or to improve the furnishing of present ones. A demonstration of such willingness occurred in July 1969 when a vigorous dispute between Comsat, on the one hand, and RCA Communications, ITT World Communications, and Western Union International, on the other hand, over the distribution of operating tasks and ownership of an earth station in Guam was resolved as follows:

Ownership: Comsat would own 50%, while the other three companies would share in proportions to be determined by the FCC.

Operating tasks: Comsat would become earth station systems manager, while RCA Communications would become earth station operations manager, responsible for such matters as procurement.

(ITT Space Communications, another subsidiary of ITT, had earlier received a contract for the construction of the station.)[24]

Experience seems to demonstrate that regulatory processes

which are aimed at eliminating, through merger or by other means, individual public utilities as separate companies often consume a great deal of time and energy on the part of both the private and the public sectors. If this is true of such companies in general, it is likely to be even more true of government-sponsored companies. Governments which sponsored such companies are likely to assume certain political and "moral" responsibilities not only for their corporate survival but also for their successful operations. Therefore, such sponsorship is likely to inject into the government's overall judgments of what constitutes acceptable performance of the communications system as a whole, a factor which may lead other companies to have less confidence in the objectivity of such judgments. This circumstance in and of itself constitutes a significant element in judging whether government-sponsored companies constitute a suitable organizational device.

Procedures and Proceedings

Like other conflicts between individuals or organizations, conflicts over the distribution of tasks and opportunities can be resolved peacefully either by agreements between contending parties (consensual procedures) or by procedures which involve the imposition of determinations by third parties which usually are governmental organizations (adjudicatory procedures*). In the course of adjudicatory proceedings, the contending parties usually are given opportunities to present proofs and reasoned arguments in support of the particular distribution which they propose, and the governmental organizations are expected to decide upon a particular distribution in accordance with some general

* The meaning of the term "adjudicatory" as used here is broader than the legal term which is used to distinguish "adjudicatory proceedings" from "rule-making proceedings." In the broader sense, rule-making proceedings designed to distribute tasks and opportunities are adjudicatory proceedings.

rules or principles, even though these may be exceedingly indefinite.

Formality in various respects constitutes an important aspect of adjudicatory processes. Such formality may take the form of requirements with regard to (1) the kinds of proofs and reasoned arguments which may be presented, (2) the manner in which they are presented, (3) the manner in which the governmental organizations consider the proofs and arguments which are presented, and (4) the manner in which they must support with reasoned arguments their decisions to adopt particular distributions of tasks and opportunities.

The degree of formality which must be followed varies according to the particular governmental organizations which determine the distribution of tasks and opportunities, and the type of proceedings in which such determinations are made. It is greatest in proceedings conducted by courts. Proceedings conducted by independent regulatory commissions are more formal when they seek to distribute tasks and opportunities between two specific contending parties than when they seek to lay down general rules applicable to the distribution of tasks and opportunities among several groups whose membership is subject to change.

In the more formal types of adjudicatory proceedings which involve a few specifically identified companies that advance mutually exclusive schemes for distributing tasks and opportunities, relevant technical information must be presented for a formal record by experts who are subject to cross-examination, and the regulatory commissions are limited to considering the formal record. The original Carterfone proceeding involving AT&T, GT&E, and Carter Electronics is an example of such a proceeding.

In somewhat less formal proceedings which involve groups of companies, a greater number of overlapping distributive schemes are likely to be presented. The relevant information may be submitted in writing for a formal record, but cross-examination of experts is likely to be considered unnecessary and the governmental organizations are not strictly limited to considering the formal record. In reaching their decisions, they may consider information other than that presented for the formal record, including supplemental information secured informally from representatives of the contending groups. The proceedings involving AT&T's new tariffs regarding foreign attachments and interconnections which were filed in the wake of the Commission's Carterfone decision are an example of such less formal proceedings.

The purpose of such less formal proceedings is to explore whether agreement can be reached, in some respects at least, on distributions of tasks and opportunities, thus obviating the necessity on the part of the governmental organization to impose an adjudicated decision on unwilling parties. Thus consensual procedures may be blended with adjudicatory ones.[25]

There is still a considerable amount of resistance both in and out of government to relying on consensual procedures. In government, there are partisans of a school of thought which holds that only formal proceedings provide the government with the club it badly needs to deal with industry, and that consensual procedures are likely to result in "deals" and "give-aways." This position was given strong expression by Commissioner Nicholas Johnson of the FCC in the course of proceedings following the initial Carterfone decision. In those proceedings, the Commission, the carriers, the hardware manufacturers, and numerous other interested governmental and private groups had to face the problem of coming up with a new regulatory scheme to deal with the conditions under which customer-owned equipment might be attached to, and private communications systems might be interconnected with, the carriers' switched telephone network. In order to arrive at a suitable new scheme, numerous complex technical and organizational features had to be balanced in order to achieve desirable degrees of flexibility and freedom of choice, on the one hand, without foregoing desirable levels of reliability and organizational responsibility, on the other hand.

A majority of the Commission felt that this objective could best be accomplished, at least initially, by means of informal technical conferences between representatives of interested parties and the Commission staff rather than formal proceedings. Commissioner Johnson dissented and advocated conducting formal hearings immediately on the grounds that:

> . . . failure to do so can only result in unnecessary delay. Informal procedures have no advantage over the formal hearing process in the situation. This is not a negotiation among sovereigns where forcing a party to take a public position may make it more difficult for him to back off gracefully. It is a matter for decision by a public body entrusted with the duty to make a decision, and with the power to enforce it. Furthermore, the Carterfone proceedings furnish ample evidence that the hearing process is an excellent means of testing technical claims.[26]

While not agreeing necessarily with Commissioner Johnson's thinking, many lawyers favor formal proceedings over informal ones for a number of reasons. They have been trained in law school primarily to be effective advocates of their clients' points of view rather than to be flexible negotiators and imaginative developers of alternative schemes for the achievement of complex objectives. Therefore, many do not feel comfortable in an informal setting which has as its purpose to explore alternatives and to narrow existing conflicts rather than to plead for particular solutions of such conflicts while eliminating from consideration all other solutions. Second, advocacy is the bread and butter of many lawyers, and therefore they have an economic stake in participating in large numbers of formal proceedings. Third, however, there is a genuine belief which many lawyers share with laymen that formal adversary procedures constitute "due process" *par excellence* if not to the exclusion of all other procedures. The limitations of adversary procedures, which are aimed at adjudicating conflicts, are perceived only dimly and very slowly. As the number of parties to such procedures increases, as their interests become more diversified, as the issues are no longer choosing between "right" and "wrong" solutions but generating solutions which reconcile multiple conflicting objectives, formal adversary procedures become increasingly unsuitable tools for the achievement of the desired objective.

Therefore, in seeking to distribute tasks and opportunities within the communications system, governmental organizations increasingly seek to avail themselves of consensual processes to supplement or take the place of formal adversary procedures. The more formal the procedures are, the more they tend to polarize the proofs and reasoned arguments advanced by the contending parties. Each party to such proceeding seeks to demonstrate that his scheme of distributing tasks and opportunities is the only "right" one while the opponents' schemes are "wrong." Privately and off-the-record, the contending parties may be willing to concede that some features of their opponents' schemes are not totally devoid of merit, and that they might be able to live with an alternative scheme which incorporates a combination of features, some of which may be drawn from the proposals which have been advanced plus some new features which may not be contained in any of them.

Governmental organizations have learned by experience that alternatives other than those formally presented by the parties for the record may be developed if the procedures followed by the governmental organizations are sufficiently exploratory and informal. The governmental organizations have also found that occasionally such alternative schemes have greater merit in protecting the public interest than the more extreme schemes originally advanced by the contending parties.

Furthermore, the agencies have learned from experience that task distributions which involve complex technical considerations lend themselves even less than distributions of opportunities to the "right-wrong" approach of formal adjudicatory proceedings. In the case of task distributions, governmental organizations frequently face the problem of how in the public interest to strike a workable balance between several desirable aspects, while at the same time reducing to an acceptable level undesirable aspects which are inextricably tied to the desirable ones.

Another aspect which governmental organizations have to consider in selecting more or less formal procedures is what priority should be assigned to a particular proceeding in relation to other proceedings with which the governmental organization is concerned and, in the light of such priority, what allocation should be made of the financial and manpower resources available. Governmental organizations may seek outside assistance in the case of proceedings concerned with highly technical problems of task distributions because the extreme scarcity of high-grade in-house technical manpower makes it necessary to husband carefully the limited supply.

In the case of the informal proceedings following the Carterfone decision referred to earlier, for example, the Commission staff sought independent technical expertise to aid it in conducting the informal sessions. On July 3, 1969, the Commission announced that the National Academy of Sciences would select an ad hoc panel to provide the needed assistance consisting of "defining and evaluating alternative solutions and giving consideration to immediate and long-range implications." [27]

The scope of the assistance to be given by an organization

such as the National Academy of Sciences presents some difficulties. While the Academy was established for the purpose of giving technical assistance in all fields of human knowledge, almost from the begining "some of the most important questions which the Academy has been asked to consider, have not related to any particular branch of science, but rather to matters of public policy." [28] The role of the Academy as a possible policy maker in the post-Carterfone proceedings was objected to by the Business Equipment Manufacturers Association.[29]

If any evidence is required to demonstrate the polycentric nature of the issues and conflicts inherent in the Carterfone complex, the list of participants in the informal proceedings presided over by the panel appointed by the National Academy of Sciences provides telling evidence. The list is as follows:

Companies

Advance Technology Systems, Inc.
Acron Corp.
American Broadcasting Co.
AMP, Inc.
Aeronautical Radio, Inc.
American Telephone & Telegraph Co.
American Telecommunications Corp. (Deco-Tel)
Ballistics Control Corp.
Bethlehem Steel Corp.
Thomas F. Carter and Carterfone Communications Corp.
Collins Radio Co.
Comtel Communications Corp.
Data Access Systems, Inc.
Computer Group, Inc.
E. I. duPont de Nemours & Co.
EDP Resources, Inc.
Ford Motor Co.
General Electric Co.
General Telephone and Electronics Corp.

International Communications Corp.
International Telephone and Telegraph Corp.
McGraw-Edison Co.
Microwave Communications, Inc.
Monsanto Co.
Motorola, Inc.
National Broadcasting Co.
National Communications & Electronics Corp.
Northrop Corp.
Olin Mathieson Chemical Corp.
Penn Central Co.
Photo Magnetic Systems, Inc.
Republic Steel Corp.
Ripley Co., Inc.
Sanders Associates, Inc.
Southern Pacific Co.
Union Carbide Corp.
United States Steel Corp.
United Telephone System
Vidcom Electronics Inc.
Weyerhaeuser Co.
Western Union Telegraph Co.
Westinghouse Electric Corp.
Xerox Corp.

Associations

Aerospace Industries Association of America
American Bankers Association
Association of Data Processing Service Organizations (Computer Time-Sharing Services)
American Petroleum Institute (Central Committee on Communications Facilities)
American Trucking Association
Air Transport Association of America
Business Equipment Manufacturers Association
Electronic Industries Association (Communication Terminals & Interfaces Section)
National Retail Merchants Association

U.S. Independent Telephone Association
Utilities Telecommunications Council

Government Agencies

Secretary of Defense
Secretary of Transportation

The procedures employed by private and quasi-public research organizations which are retained by governmental organizations in such situations are likely to be similar to the consultative, exploratory procedures followed by such research organizations when performing work for nongovernmental clients. The retention by the Federal Communications Commission of the Stanford Research Institute in the case of the Commission's computer proceeding is an example. In addition to analyzing the formal briefs submitted by the numerous contending groups and individual companies, Stanford Research Institute conducted informal interviews with such groups and companies with a view of developing additional information which might be helpful in determining alternative approaches to the regulatory problems at hand.

If the distribution of tasks which involve technical and managerial considerations makes a resort to informal consensual proceedings frequently advisable, the distribution of opportunities calls similarly for proceedings of this type, but for somewhat different reasons. The distribution of opportunities deals with the question "who gets what and how," and that question has been recognized as being an eminently political question. It is commonly accepted by students of political processes in the United States that elected political bodies and elected political officials in deciding political questions rely extensively on negotiating and bargaining procedures. It is not uncommon in recent history for conflicts over the distribution of opportunities within the communications industry to be considered by Congress and state legislatures,

Presidents, governors, and other elected officials. Comsat, pay-TV, and cable television legislation are examples.

It is not equally accepted, however, that conflicts over opportunity distributions remain essentially political conflicts even though resolutions of such conflicts are delegated to executive departments or independent regulatory commissions to be decided in accordance with such general standards as "the public interest." In other words, the political nature of these conflicts remains, in spite of frequent expectations that such conflicts "should" be decided "objectively" by neutral experts or technicians in accordance with principles of law and economics.[30]

Under these circumstances it is only realistic to expect executive departments or independent regulatory agencies charged with the responsibility of resolving conflicts to resort to informal negotiating and bargaining procedures. Not infrequently, the parties may be urged to conduct such negotiations on their own and to present any resulting agreements to the departments or agencies for approval. Such urging may be accompanied by thinly veiled warnings that the parties may much prefer to distribute opportunities in accordance with an agreed-upon formula than to have imposed upon them a distribution formula which they find utterly unpalatable. For example, Rep. Torbert H. Macdonald, Chairman, Subcommittee on Communications and Power, Committee on Interstate and Foreign Commerce, House of Representatives, in speaking at the Convention of the National Cable Television Association in San Francisco in June 1969 stated:

> I doubt that some Members of Congress will be overly tolerant of sharp divisions after we have had assurances from responsible spokesmen for your industry and others that you have reached accord. By the same token, dissent among the leaders of the other segments of the industry involved with this manysided problem will be watched closely to see that

> constructive efforts for a solution do not take a back seat to just plain subbornness. I salute Mr. Wasilewski [President, National Association of Broadcasters] and Mr. Ford [President, National Cable Television Association] for their efforts to achieve realistic solutions. Congress is not as gifted in solving these problems as people in the industry.

In order to protect the public interest as well as the interests of parties which may be affected adversely by governmental decisions reached on the basis of informal proceedings, safeguards comparable in some respects to those which have been developed for formal administrative proceedings are necessary.[31] In the case of informal intercompany and government-industry conferences, the Department of Justice has insisted upon "adequate representation of divergent interests" and "wide circulation" of any proposals resulting from such conferences.[32] Going beyond the Department's requirements, there is no reason why such decisions, before becoming effective, should not be made public and be accompanied by a statement of reasons why such decisions are believed to be in the public interest. Such procedures would give interested parties an opportunity to submit information and to state reasons why, in their opinion, such decisions should not be permitted to take effect.

Procedural safeguards were developed to legitimize the more formal administrative proceedings as we know them today. A similar evolutionary process can be expected to take place with respect to informal administrative proceedings which are likely to be used increasingly in the future.

CHAPTER V

The Future Regulatory Environment and Appropriate Industry Responses

What will the future regulatory environment of the communications industry be like, and in what respects are the structure-determining processes likely to differ from those which have determined the present industry structure? What roles might individual companies and the industry as a whole play in such future processes? These are the questions to which we shall address ourselves in the subsequent pages.

Future Regulatory Environment

Organizations and Procedures

If present trends continue unabated, the communications industry, like other regulated U.S. industries, will operate in a political economy in which "macro-politics" will become increasingly influential. This terminology means that political processes on the highest national levels involving the President, Congressional leaders, and the national community will become increasingly implicated, in addition to lower level political processes and processes of economics.[1] This escalation to the macro-political level will occur for several reasons: (1) problems involving the communications industry will have an increasing impact on other national and international social, political, and economic problems, and the reverse will be equally true; (2) as the industry expands and companies become more differentiated, the intermediate political processes will increasingly lack the resources and political standing necessary to resolve major conflicts and to provide the overall guidance and coordination required to assure that future communications systems will meet diverse domestic and international needs; and (3) individual companies, industry segments, and other groups that are disappointed in their expectations or that seek to defend the status quo will insist increasingly on appealing their cases to the macro-political level for the purpose of securing reversals of decisions reached on the intermediate political level.

For these reasons, it can be expected that responsibilities will be assumed increasingly by political leaders in the executive branch on the macro-political level with regard to (a) establishing future communication objectives and coordinating programs designed to achieve those objectives, and (b) resolving major conflicts between companies, industry segments, and other interested groups with regard to such objectives and programs.

Supporting evidence for these conclusions may be found in recent historical events where macro-political levels became involved in resolving major conflicts and developing basic policies in connection with U.S. international satellite communications. The latest example of escalation to the highest macro-political level of satellite communications policy questions may be found in the formation of a small working group in the White House for the purpose of formulating "Administration suggestions or comments" regarding the introduction of communication satellites into U.S. domestic communications. The White House announcement, dated July 22, 1969, stressed: "We will be concerned, of course, with the general structure and direction of the industry and not with specific applications pending before the Commission." The FCC was invited "to participate in any way you deem appropriate." According to reports in the trade press, the first meeting of the working group included, in addition to special presidential staff assistant, Clay T. Whitehead, representatives from the Office of Telecommunications Management, the Office of Science and Technology, the Council of Economic Advisers, the Bureau of the Budget, the Post Office, Justice, and Transportation Departments, the National Aeronautics and Space Administration, and the Federal Communications Commission.[2]

On October 15, 1969, speaking for the three television networks, President Frank Stanton of CBS expressed gratification that President Nixon had taken a special interest in the subject of domestic communication satellites. He expressed the hope that an early decision would be reached on a proposal made public at that time calling for the formation of a consortium by the three networks to operate a special purpose domestic satellite system which would make the networks independent of AT&T's landlines and microwave facilities. AT&T had raised the rates which it expected to charge the networks for interconnection services from $45 million to $65 million annually. The proposed satellite system would cost approximately $100 million and would provide live network broadcasting to Alaska, Hawaii, Puerto Rico, and the Virgin Islands which now have only delayed programming. In addition, the proposal contemplates that the Corporation for Public Broadcasting would become the fourth consortium member and that channels of the system would be made available for public (educational) broadcasting free of charge.

AT&T commented that "the wisest public policy at this time would be to permit any organization or group interested in establishing a domestic satellite system—including the networks—to apply for a license to establish and operate such a system. We believe that this approach would offer flexibility and incentive for creative private initiative, and would provide the most appropriate means for an orderly development of domestic satellites. . . . Our recent studies indicate that satellite costs currently may be less favorable compared to terrestrial costs than appeared to be the case some years ago.

. . . Looking to the future, AT&T anticipates that when it makes good technical and economic sense to do so, it will seek authorization to use satellites in its own operations."

James McCormack, Chairman of Comsat, countered with a proposal under which Comsat would provide the services required by the television networks as well as services which might be sought by press associations, cable television systems, and other industrial users. It was announced that some of the top managers of Comsat, CBS, NBC, and ABC, and the Corporation for Public Broadcasting would meet to discuss the use of satellites to relay television programs. It can be anticipated that meetings with other groups interested in domestic satellite relay and other services will follow so that the various uses which might be made of the system may be explored.

There is further likelihood of increased macro-political participation because the intermediate political level apparently is unable to resolve major conflicts between cable television operators and broadcasters, on the one hand, and telephone companies, on the other hand, and to establish objectives and coordinate programs for novel mass communication and intercommunication services. Similarly, the rapidly evolving conflicts between the communications industry, which transports information, and the computer industry, which stores, retrieves, and processes information, and the need for establishing objectives and coordinating programs involving those heretofore separate industries, are likely to involve the executive branch on the macro-political level.

Evidence is also available in the form of recommendations contained in various studies of communication policy development and coordination which were made for Presidents Truman, Eisenhower, Kennedy, and Johnson. These studies have recommended unanimously that a capability for developing and coordinating communication objectives and policies, and for resolving major group conflicts, be created on the macro-political level of the Executive Branch. The recommendations have differed, however, with regard to where such capability should be placed.

The conviction that such a capability is required has become stronger as the years have gone by and as different groups of experts have sought to come to grips with the problem of developing and coordinating communication policies. Initially, the experts thought that it should be possible to develop some overall policy which might be enacted into law, or to issue a series of policy statements which might set forth authoritatively U.S. policy with regard to a number of communication issues. The latter approach sought to emulate the adoption of a policy statement with regard to U.S. international aviation, which favored competition among a limited number of U.S. international air carriers rather than reliance upon a single chosen instrument. The latter concept, which the policy statement rejected, had been advocated in particular by Pan American Airways.

With regard to the development of a general U.S. communication policy, it soon became apparent that the objectives of such a policy could not be stated with much greater specificity than was achieved in Section 1 of the Communications Act of 1934, which defines the objectives as follows:

> . . . to make available, so far as possible, to all the people of the United States a rapid, efficient, nation-wide, and world-wide wire and radio communication service with adequate facilities at reasonable charges, . . .

The initial study by a Presidential Board [3] stated a theme which subsequent studies have reiterated with more or less emphasis. The theme was that in order to achieve the objectives stated in Section 1 of the act, the policies adopted would have to be two-pronged: (1) for the government to encourage and promote the health of those privately owned companies which play the principal part in the country's telecommunications system and which, therefore, constitute a vital national asset; and (2) for the government to remain alert to the problems of the telecommunications system as a whole and to be prepared to adopt measures necessary to insure the continued satisfactory operation of that system. The study, therefore, concluded that there would have to be some capability on the macro-political level to implement those policies.

A subsequent study stressed the "unfortunate absence" on the macro-political level of "an organization adequate to accumulate the information and experience upon which to base sound action" and "with adequate knowledge, experience and stature to act for the President on these matters." [4]

Efforts to create a capability on the macro-political level to assist and act for the President began in 1951 with the appointment by President Truman of a Communications Advisor to the President. In 1953 President Eisenhower transferred the office of the Advisor to the Office of Defense Mobilization (ODM), and later that year the position of Assistant Director for Telecommunications was created in ODM. This arrangement was modified in 1962 by President Kennedy.[5] The arrangement then made has continued in effect with minor modifications until the present. It is as follows: the Director of Telecommunications Management is an Assistant Director of the Office of Emergency Preparedness. He has general coordinating responsibilities in formulating government policies. In addition, he is a Special Assistant to the President for Telecommunications, and in that capacity he is expected to advise the President on telecommunication matters. The Director of Telecommunications Management also discharges responsibilities with regard to (1) assigning frequencies to government agencies, (2) promoting and coordinating satellite communication activities, and (3) formulating policies for the National Communications System, which was established to coordinate the operation of the domestic and international communication facilities of the Federal government.

The adequacy of the Federal government's leadership in

coordinating the development of the communication system as a whole is currently again under study. It is contended by some well placed individuals in the Nixon Administration that the present organization is inadequate primarily because the Director of Telecommunications Management lacks the resources, prestige, and political backing which are indispensable to exercising effective leadership. They argue that, being narrowly specialized and, therefore, having infrequent access to the President, the Director's leadership potential in relation to industry, other government agencies, and Congress is not likely to be improved substantially even if his office were taken out of the Office of Emergency Preparedness and set up as a separate office in the Executive Branch of the President as has been suggested by some. They contend that the resources, prestige, and political backing necessary to exercising adequate leadership are not likely to become available unless the leadership responsibilities are vested in a member of the President's cabinet. In other words, one of the present departments (such as the Department of Transportation or the Department of Commerce) or, less likely, a new Department of Communications should be called upon to discharge these responsibilities.

Several reasons are advanced why such leadership responsibilities should be reposed in one of the departments and not in the Federal Communications Commission. The reasons pertain to the nature of the responsibilities as well as the nature of the procedures which must be employed if these responsibilities are to be discharged effectively. Responsibilities relating to the coordination of telecommunication programs should not be separated from Presidential responsibilities relating to the national economy, the national security, and the entire national social and political fabric. All of these responsibilities are political and managerial by nature, and must be exercised with the aid of experts who are skilled in various disciplines and who are accustomed to cooperate with each other in solving problems in broad areas where public policies must be formulated and coordinated. Since the supply of such experts is exceedingly limited, and since it is difficult to attract and keep such personnel, it is imperative, so it is argued, that the leadership responsibilities be placed in a department because only a department which is presided over by a member of the President's cabinet has the necessary prestige to attract and keep such personnel.

The second line of arguments suggests that the procedures employed by a department, in contrast to the procedures followed by an independent regulatory agency, are more suitable to the discharge of broad political-managerial responsibilities. An agency, it is argued, is more apt to conduct proceedings which deal separately with various problems although the problems dealt with are interdependent. The FCC, for example, at present conducts separate proceedings on a number of interrelated problems such as cable television, domestic satellites, computers and communications, specialized common carriers for business communications services, and bulk rates charged by regular carriers for business communications services and facilities. There is a considerable overlap with respect to the companies which are parties to those proceedings, and by dealing separately with interrelated problems, the Commission cannot take advantage, in attempting to resolve conflicts among those companies, of trade-offs which might be developed between those various problem areas. Developing and taking advantage of those trade-offs necessitates, among other things, negotiating, consulting, prodding, conciliating, exploring alternative solutions of interdisciplinary problems, and stimulating experiments on a trial and error basis.[6] If such activities, which necessitate the use of informal procedures, were undertaken by independent regulatory agencies this would seriously prejudice the proper exercise of their roles as arbiters in the more formal adversary-type proceedings.

A decade or two ago, some of the regulatory procedures which were employed by independent regulatory agencies were criticized on the grounds that they acted as prosecutors, judges, and juries. Should these same agencies now attempt to act in addition as brokers, negotiators, and prodders in attempting to coordinate industry activities, the community's sense of justice and fair play would almost certainly be violated. Very appropriately, therefore, it is pointed out, a division of responsibilities has been achieved in the field of transportation in which the resolution of conflicts requiring more formal procedures is left, as before, to the Civil Aeronautics Board and the Interstate Commerce Commission, while general coordinating and policy-making responsibilities, which often are exercised on the basis of informal negotiations and explorations, have been vested in the Department of Transportation. If the Department seeks to influence the outcome of individual formal proceedings before the CAB or ICC, it does so by intervening in such proceedings, where it then may file briefs or present witnesses as do other parties to such proceedings.

Another line of arguments pertains to the relationship between antitrust enforcement and communications policy development and coordination. It is argued that traditional antitrust concepts focus on performance of portions of systems but that in developing and coordinating communication policies on a system-wide basis, antitrust concepts have to be modified. Modifications are facilitated if a department which is responsible for policy development and coordination is able to deal with the Department of Justice, which is responsible for antitrust enforcement.

Persons who advocated placing responsibilities for policy development and coordination in one of the present or a new executive department realize the drawbacks inherent in asking one department to coordinate communication functions of other departments, especially if such a department has a built-in bias with regard to its own communication functions. These persons expect to overcome such difficulties through appropriate procedural safeguards, including review panels appointed by the President in cases of unresolved interdepartmental conflicts.

The macro-political level in the United States is not unique in seeking to achieve a greater capability for communication policy development and coordination. The Canadian government recently established a Department of Communications (DOC) headed by a cabinet minister. In establishing the Department, transferring to it existing capabilities, and defining its responsibilities in relation to industry and other government departments and agencies, the Canadian government followed reasoning substantially identical with the reasoning described in the preceding pages advanced on the macro-political level in the United States. An additional factor contributing to the establishment of the Department of Communications was the decision to create a Canadian corporation for domestic telecommunications by satellite (Telesat Canada). The corporation is designed to achieve the dual objective of providing better communications (including television) for the northern parts of Canada and promoting bi-lingual capabilities by broadcasting television programs in English and French throughout the entire nation. Ownership in Telesat Canada will be shared by the government, the general public, and established communication companies in proportions to be determined by the government.

The Canadian government transferred to the DOC technical communication capabilities which formerly were placed in other departments. DOC was given responsibility for managing the radio spectrum and government communication facilities. Broad powers were given to DOC to

> . . . coordinate, promote and recommend national policies and programs with respect to communication services for Canada, including the Canada Post Office; [to] promote the

> establishment, development and efficiency of communications systems and facilities for Canada; [to] assist Canadian communication systems and facilities to adjust to changing domestic and international conditions; . . . [and to] compile and keep up to date detailed information in respect to communication systems and facilities, and of trends and developments in Canada and abroad relating to communication matters. . . .

The powers of the Commission on Transport, which has regulatory powers over transportation and communication companies similar to but less extensive than those exercised by the FCC, were left intact, but DOC may intervene in proceedings conducted by the Commission.

How macro-political processes may be used to affect the structure of the Canadian communications industry in important respects is exemplified by the recent enactment of legislation with regard to the Bell Telephone Company of Canada.

Since Bell Canada is a "special act company," changes in the company's charter require legislative action by the House of Commons of Canada. The company, which is 95% Canadian owned, sought legislation (1) to increase its authorized capital from $1 billion to $1.75 billion, (2) to broaden its authority from engaging in telephone operations to engaging in the business of providing electromagnetic telecommunication services or facilities "using and adapting any improvement or invention or any other means of communicating," and (3) to clarify the company's legal authority to own Northern Electric, a manufacturing subsidiary which plays a role comparable to that of Western Electric in relation to AT&T.

In the course of hearings on the legislation held by the Standing Committee on Transport and Communications, several independent hardware manufacturers and a firm of consultants specializing in computer-communication services objected to Bell Canada's policies relating to in-house hardware procurement and the attachment and interconnection respectively of customer-owned equipment and private communication systems. On the basis of policies formulated on the macro-political level and consultations which ensued between the committee chairman, representatives of Bell Canada, and Canadian officials responsible respectively for communications regulations and antitrust enforcement, amendments to the legislation were formulated which were adopted subsequently by the House of Commons. These (1) preclude further acquisitions by Bell Canada of communications hardware manufacturing companies but permit acquisitions of research and development companies; (2) permit attachments and interconnections in accordance with reasonable requirements prescribed by Bell Canada, (3) authorize interested parties to secure (a) a regulatory agency review of those requirements by the Canadian Transport Commission and (b) judicial review of the Commission's decisions.

By means of a single piece of legislation, conflicts were resolved with respect to the future distribution of communication tasks and opportunities between (1) Bell Canada, (2) independent hardware manufacturers, (3) companies which might seek to attach to the telephone network equipment owned by them, or to interconnect with the network private communications systems owned and operated by them, (4) broadcasters, and (5) cable television operators. As a result of the utilization of macro-political processes, Bell Canada is now authorized to furnish communication services of *any* kind, except broadcasting and CATV, utilizing for such services *any* technology which might become available. The "official explanatory notes" which accompany the statute contain the following explanation of Bell Canada's broadened charter:

> The revolution in communication techniques demonstrated that the company can no longer be considered exclusively as a telephone company. In order to remain strong

> and competitive and thus be an asset to the Canadian economy, it is compelled to meet the demands of Canadians and to supply them with the widest possible range of telecommunication services.

Furthermore, Bell Canada's continued ownership of Northern Electric was legitimized, but Bell Canada was required to forego expanding by merger or acquisition its hardware manufacturing capabilities. Additionally, the company had to relinquish some of the exclusivity of its future hardware business opportunities and will have to face in the future increased competition from private communications systems. On the other side of the ledger, future tasks and opportunities of hardware manufacturers have been expanded, and the tasks and opportunities of broadcasters and cable television operators have been protected.

Proposals aimed at strengthening the U.S. government's macro-level capability to a point where it may exercise effective leadership in identifying and establishing communication objectives, coordinating programs aimed at achieving those objectives, and resolving major intercompany and interindustry conflicts with regard to such programs may temporarily have a relatively low priority considering the urgent need for solving other domestic and international problems which confront the Federal government. While there is at present no acute emergency which would require immediate action on such proposals, there is an increasing likelihood that such proposals will be adopted. There is a growing apprehension on the part of well-placed individuals in industry and government that unless such action is taken reasonably soon, the introduction of new communication services or the use of new modes of transmission to improve existing services will be greatly delayed. This is likely because large companies, other than the present carriers, which might be capable of participating in new ventures designed to furnish new communication services by means of new technologies, are likely to decide, in the exercise of prudent business judgment, that the prospects of pursuing other opportunities are more inviting. Such conclusions are likely to be reached because the offering of new communication services would pit such companies against present carriers, and would involve them in fighting battles on all levels and before all branches of government.*

Antitrust involvement on the part of such major companies as, for example, IBM, in connection with their present business activities is also likely to act as a substantial constraint on any plans which such companies might consider in the field of communications. This then leaves the planning of new services, or of existing services by alternative methods, either to the existing carriers or to companies which have fewer commitments, fewer opportunities, fewer constraints, and fewer resources. Microwave Communications and Carter Electronics Coroporation are examples of such companies. Some independent telephone companies, while expressing the view that liberalized attachment and interconnection policies were "ideas whose time had come," did not take the initiative in putting such policies into effect but let Carter and his attorneys battle the issues out in the courts and before the FCC. The government, however, will continue to be reluctant to let existing carriers, and especially AT&T, expand their "monopoly" positions.

The existence of these basic policy conflicts renders all the more likely the adoption of proposals designed to strengthen the government's capability on the macro-political level to exercise leadership in resolving intercompany and interindustry conflicts, and in guiding the development of the future structure of the communications industry. Even if such proposals should be adopted and one of the executive departments be designated to discharge those responsibilities, this

* An example of such a situation is presented in Appendix D.

would not necessarily preclude direct involvement on particularly important occasions of the President himself, especially if several cabinet officers should find themselves in substantial policy disagreement. For example, the question of what companies should be authorized to furnish domestic communication satellite services would still be likely to lead to an escalation, as it has at present, to the highest macro-political level — namely, the President himself.

New Framework for Reasoned Arguments

If structure-determining decisions are escalated to the macro-political levels, what is likely to be the intellectual framework within which such decisions will be reached, and in what respects is it likely to differ from the present framework? In making these decisions, the leaders at the macro-political level are likely to consider (1) demands advanced by groups concerned with the decisions, (2) alternative "rational" distributions of tasks and opportunities which their expert departmental staffs have recommended, and (3) if pertinent to such distributions, the fears and expectations of the community at large as interpreted by the President and his immediate associates.

Distributions will be considered "rational" if they are designed to achieve, in connection with the furnishing of specified ultimate communication services, desired long-range performance goals of the system as a whole and if such distributions are based on analyses undertaken with the aid of all available intellectual tools.

The intellectual framework which will be employed, therefore, will seek to encompass (1) ways of measuring the comparative economic and political club of the numerous and diverse groups which will advance conflicting claims, (2) the multiple and often conflicting objectives which the communications system is designed to accomplish, and (3) the conflicting reasoned arguments which will be advanced by various disciplines to accomplish rational task and opportunity distributions within the communications system.

A framework which meets such diverse needs will be more pragmatic and less dogmatic, because it will be influenced by the thinking of politicians as well as their chiefs of staff. These increasingly will be individuals who will have learned to think in terms of very large systems, such as systems designed for space exploration and utilization, systems which might improve the livability of our megalopoli, systems for moving goods and people both between and within megalopoli, and systems for transmitting, distributing, storing, retrieving, and processing information.

Such a framework will seek to focus more on the demands for, and the technical feasibility and costs of, *future* communication and information services, and their relations to each other and to existing services, and less on *historical* costs of particular services and histories of performance of particular parts of the communications system. Going beyond economic data, such a framework will include social and political factors. Therefore, such a framework will require a much broader information base, suitable for evaluating alternative programs in economic, social, and political terms. There is a general tendency to turn to economists to undertake such evaluations in the expectation that the tools they employ for economic analysis will make possible precise evaluations of public policy alternatives. As Douglas Needham has pointed out, however, economic analysis is not designed to make social or political choices but only to assess the economic costs of alternative public policies:

> Which structure of industry is desirable depends upon the nature of the objectives being pursued. Some objectives may have little to do with economics. . . . The pursuit of some objectives may even conflict with the achievement of economic objectives. . . . The economist has no special competence for choosing what the objectives of public policy

> shall be. Nonetheless, it may still be important to know the cost, in terms of alternative economic benefits foregone, of achieving non-economic objectives.[7]

Such a broader framework will make use to a greater extent of "continua" and other relative concepts which stress, matters of degree, the relatedness of problems, trade-offs among multiple relevant factors, and the pros and cons of various alternative solutions. Such use of relative concepts will contrast with traditional more absolute concepts which tend toward separating interrelated problems, polarizing points of view, and considering either-or solutions.[8]

Such a framework will be designed to pursue a problem-solving approach with regard to the distribution of tasks and opportunities, and will lead to readier acceptance of the fact that each solution will generate new problems. Therefore, it will differ from the present framework which is designed to search for solutions which are "right" as a matter of principle and which, therefore, should stand regardless of changing conditions.

Appropriate Industry Responses

If in the preceding pages we have presented a reasonably accurate diagnosis of the situation which communication companies are likely to face, individually and collectively, then the question is what would be appropriate company and industry responses to such developments.

The escalation of structure-determining processes to the macro-political level will be readily accepted or even welcomed by those companies and industry associations which have on occasion promoted such escalation to further their expectations or to protect the status quo. Other companies and associations may see such escalation as a threat rather than an opportunity. Whether they will see it as one or the other will depend largely on how well prepared and equipped they are to participate in structure-determining processes on the macro-political level.

Such preparation will require the management teams of the companies and associations to focus clearly on the twin objectives of these structure-determining processes on the macro-political level: (1) to permit retention, on an ongoing basis, by individual companies and associations of as much initiative as possible in performing particular communication tasks and exploiting opportunities connected therewith, and (2) to provide as much coordination among such tasks and opportunities as will be necessary to assure that the system as a whole will perform in an acceptable manner (by meeting general public and special private needs as well as general and special governmental needs). In other words, corporations which perform tasks within large systems must serve two masters: the system and the corporation itself.[9]

In principle, the objectives of individual companies and industry associations may be the same as those of the government. However, in developing specific policies designed to achieve these objectives major conflicts are bound to develop. Company A's desire, for example, to furnish particular new communication services and to exploit opportunities connected therewith is likely to be met by Company B's desire to prevent the execution of such plans. Company B, therefore, is likely to appeal to the government to stop Company A from executing its plans. If Company A's plans and Company B's opposition are not of major significance for the functioning of the system as a whole, the conflict may be resolved on the intermediate political level. If, however, Company A is only one of many companies which are similarly situated, then the conflict is likely to be escalated to the macro-political level.

Similarly, conflicts are likely to arise over the desirability of providing for coordination in particular situations. Should Company A be required to establish and publish standards for equipment which it manufactures for use in providing communication services, so that Companies B, C, and D will be in a position to sell to Company A a portion of the total amount of the equipment required by Company A? If Company A refuses, and there are many Companies B, C, and D, the conflict is likely to be escalated.

Disagreement may arise also over whether such standards should be performance standards or equipment standards, and whether equipment different from that manufactured by Company A should be accepted by Company A as long as it meets the performance standards and is compatible with Company A's equipment. In such case, the macro-political level may be petitioned by Companies B, C, and D to establish standard setting procedures.

Individual companies and industry associations, therefore, will have to seek answers to lines of questions which most of them have not been in the habit of asking: (1) What will be the nature of tomorrow's system and how is it likely to differ from today's? (2) What tasks and opportunities do we want to pursue within the future system in addition to, or in lieu of, present tasks and opportunities? (3) Will our plans dovetail with those of other companies, or may ours or theirs have to be modified to achieve the necessary degree of coordination within the system? (4) If intercompany or interassociation conflicts over such plans are not resolved by companies or associations, what is the outcome likely to be if government on the macro-political level becomes involved in resolving such conflicts?

New Framework for Reasoned Arguments

In order to be better prepared to answer such questions, company and association managers must be prepared to think and act within an intellectual framework which permits them to generate alternatives to policies which officials on the macro-political level are prepared to adopt and execute in the absence of acceptable alternatives. Such a framework should serve as a constant reminder to company and association managers that there are numerous ways in which different government agencies can seek to accomplish various governmental communication objectives. For example, government agencies may provide communication services for their own use as in the case of military communication services. They may provide such services for the benefit of the general public as was done until recently in Alaska. Alternatively, government agencies may look to private companies to provide, directly or indirectly, services for governmental or general public uses. Toward this end, government agencies may establish new companies, as they did in the case of Comsat, or they may issue franchises to one or several companies. In order to achieve fair competition between companies, government agencies may require such companies to establish subsidiaries to furnish specific services, or they may seek to accomplish such objective by prescribing terms and conditions under which such services may be provided. Instead of regulating companies directly, government agencies may foster schemes of industry self-regulation. They may proceed to do these things in a very formal manner by enacting legislation, adopting commission rules, or bringing antitrust suits, or they may seek to provide guidance by informal suasion.

In order to be able to respond effectively to such a wide range of governmental approaches, company and association managers must be proficient in all of them and be capable of shifting rapidly from one approach to another. Better yet, they may want to be capable of anticipating governmental intervention and to take the initiative in developing coordinating plans of their own, thus attempting to make government agencies respond to company or association initiative rather than the reverse. Being properly equipped in these respects means also, however, that the management teams of the companies and associations must be capable of interrelating and integrating into consistent plans relevant technical, economic, and political information. To get individuals trained to think in terms of different goals and to employ different concepts and different thought processes in working toward those goals, to work together in developing such plans, presents a major management challenge.* Some company lawyers, for example, who are in charge of handling their companies' proceedings before the FCC, are in the habit of thinking of such proceedings as so many separate battles with other companies, each of which they are out to win. They measure their importance to the companies in terms of the number of battles in which they have participated instead of the contributions which they were able to make to the development of consistent plans to deal with their companies' external affairs. Such plans require assigning priorities to different company goals within the communications systems and developing overall strategies in dealing with other companies and government agencies in order to achieve such goals.

The disadvantages inherent in fighting separate battles instead of developing overall strategies are exemplified by AT&T's dilemma resulting from filing contradictory statements in the Microwave Communications proceeding and in support of its series 11,000 "pipe" tariff. In this proceeding AT&T had argued that there was no need for the sharing by customers of services as contemplated by the Microwave proposal. In support of its "pipe" tariff, AT&T had stressed that the tariff would permit the sharing of services by several users. The FCC majority, in its Microwave proceeding decision of August 14, 1969, stated:

> . . . we cannot ignore statements made by a party in filings

---

* Since words dominate our thinking to a considerable extent, such individuals may find it useful, in performing their integrative and interpretative responsibilities, to use words which do not have ideological connotations as monopoly and competition do. A suggested framework based on the word "apportioning" is described in Appendix C.

with the Commission which contradict or are inconsistent with the position taken by a party in an adjudicatory proceeding. The statements by AT&T in support of its own proposals substantially undermine the arguments advanced by the carrier in this proceeding to the effect that no public need exists for the sharing provisions of the Microwave proposal and that Microwave Communications is "cream-skimming." To the extent necessary to demonstrate the different positions taken by AT&T, we shall take official notice of the statements.

Organizations and Procedures

Intracompany — It is incumbent then upon top managements to organize their immediate staffs so that strategic plans will be developed and the companies' affairs will be conducted with a view to achieving the goals set forth in those plans. (In the case of multibusiness companies (conglomerates) this applies to the managements of the departments or subsidiaries which are engaged exclusively or primarily in communication activities.) The organizational problems of top managements in this respect are comparable with the organizational problems faced by the President of the United States. With the increasing interdependence of nations in today's world, planning the nation's foreign and security affairs has become of necessity a direct responsibility of the President and his immediate assistants. Therefore, recent Presidents have found it necessary, as will top managers of companies, to assign to particular individuals in their immediate staff special responsibilities for working with individuals throughout these companies who have competence in different disciplines, with a view to assembling relevant information, establishing a hierarchy of goals, and suggesting alternative policies to achieve those goals.* It is then up to top management to

---

* It is interesting to note that on July 11, 1969, AT&T announced the establishment of a Department of Environmental Affairs to be headed by one

select their goals and formulate their strategies. In moving in this direction, top managements of some companies have found it useful from time to time to rely on outside assistance in the form of consultants or advisory groups, or to invite outsiders who have demonstrated special capabilities in directing the development of integrated plans to join the top management team. Additionally, however, managements should be able to look to industry associations to assist them in discharging their greatly broadened responsibilities.

Intercompany and Industry-Government Problems — Industry associations are expected to deal with intercompany and industry-government problems. In the area of telecommunications, there is no single industry association whose charter directs it to deal with these problems on a system-wide basis. Lately, the Electronic Industries Association has made some efforts to fill this void by forming, within its Board of Directors, a Committee on Telecommunications Policy. The association, however, is concerned with intercompany and industry-government problems which encompass all of the electronics industry and which, therefore, transcend the area of telecommunications. On the other hand, the association has represented in the past primarily the point of view of manufacturers of electronic hardware. Therefore, its efforts, while pointing up the importance of exercising industry initiative in dealing with telecommunication problems on a system-wide basis, are likely to fall short of meeting the integrative needs of the various industry segments and companies in the communications industry.

---

of the senior officers of that company. Commenting on the new organization, AT&T Chairman H. I. Romnes said: "Our commitment to a continually improving communications service requires that our planning and operations be effectively related to the environment in which our service is provided and used. The new department, under Walter Straley, will provide leadership in all areas of environmental concern and act as an information clearing house and coordinator of these matters."

Traditionally, industry associations have been formed to represent the common interests of their member companies. As member companies become increasingly differentiated, the associations find it more and more difficult to formulate positions which are supported by the entire membership. As factions develop within the membership, the ability of associations to speak with a single voice on behalf of their members in dealing with government is weakened correspondingly.

As new associations are formed to represent diverse special interests, the discrepancy between the "public interest" (i.e., the balance of the interests of all "publics") and the special interests represented by various associations becomes increasingly apparent. Therefore, the credibility of the associations, that is, their reputation for trying to deal fairly with the public interest, is reduced correspondingly.

This then raises a basic question, whether the increasing differentiation of companies and associations does not call for integrative mechanisms and integrative efforts different from those provided traditionally by industry associations.

The nature of the problems confronting companies and industry segments which are parts of a regulated system was recognized by some perceptive individuals who were intimately familiar with the problems of the atomic industry. Their insights led to the formation, in 1952, of the Atomic Industrial Forum.

Why, however, should the atomic industry and the Forum be considered appropriate models for the communications and data processing industries? The reason is simply this: Technological developments in these industries have resulted in their becoming increasingly differentiated. Public policies developed at times when technologies were fewer and less advanced, and when the industries were less differentiated, no longer fit the changed conditions. The policy vacuum has to be filled in some way. If the industries do not participate actively and constructively in the policy formation process, then more or less arbitrary policies are likely to be imposed upon them by government, to the likely detriment of the industries as well as the public simply because the government has to reach policy decisions without an adequate information base.

In order to avoid the imposition by government of unsatisfactory public policies in the area of atomic energy, Dr. T. Keith Glennan, President of Case Institute of Technology, after serving for several years as a member of the Atomic Energy Commission, suggested the creation of the Atomic Industrial Forum. The Forum was established following a series of meetings attended by Dr. Glennan, interested industrial and government leaders in the field of atomic energy (including particularly Walker L. Cisler, Chairman of the Board, The Detroit Edison Company, and Gordon Dean, the then Chairman of the Atomic Energy Commission), and two experts in public administration and law familiar with problems (including antitrust problems) encountered in cooperative intercompany and industry-government relations (respectively Oliver Townsend, now Chairman, New York State Atomic and Space Development Authority, and Oscar M. Ruebhausen, a New York lawyer who was counsel to Dr. Vannevar Bush during World War II when Dr. Bush was in charge of governmental research and development programs).

The Forum is not a trade association because it does not purport to protect the special interests of any one segment of the industry and because it does not act as spokesman for any segment either in Washington or elsewhere. The purpose of the Forum is to serve industry, government, and the public on an on-going basis by studying changing policy problems, by developing responsible views on alternative solutions which are feasible to deal with such problems, and it is hoped by anticipating some of those problems. In discharging these functions, the Forum is concerned with changing technologies not primarily because of the engineering and scientific problems which such technologies present, but because of the managerial and policy implications which they have for industry, government, and the public.

The Forum philosophy acknowledges that alternative solutions may be developed which may each have certain advantages and certain disadvantages. No special efforts are made to insist upon the formulation of unified industry positions. Therefore, by comparison with the difficulties often experienced by trade associations, there is a very much reduced risk of disagreements splitting the organization down the middle to the point where its usefulness as an industry-government relations medium is seriously impaired.

In creating the Forum, the atomic industry signified its intention to exercise initiative in formulating and expressing responsible views on alternative solutions. Such views have constituted important and constructive inputs into government efforts directed toward developing workable regulatory concepts, organizations, and procedures.

The membership of the Forum consists of (1) business and professional firms likely to be engaged in some commercial phase of the development or utilization of atomic energy (Organization Members); (2) governmental and nonprofit organizations, including labor unions (Associate Organization Members); (3) individuals who are officers or employees of Organization and Associate Organization Members (Participating Members), and (4) individuals who are interested in the development and utilization of atomic energy but who do not qualify as Participating Members.

Activities of the Forum are centered around studies conducted by its committees, subcommittees, and working groups. Every effort is made to secure representation on such committees or groups from various segments of the industry and other interest groups. New study projects are launched in response to recommendations by Forum members or at the suggestion of the Atomic Energy Commission and other government agencies. Forum staff members are assigned to the groups to undertake research, coordination, and drafting of reports in connection with such projects. To deal with some issues, the Forum may convene panels or seminars or employ other suitable formats for the development of responsible views on alternative solutions of problems. When presenting views or reports to the government, the members of the committee or groups *present such views as their own.* They do *not* purport to speak for the Atomic Industrial Forum.

In addition to studies the Forum conducts an Annual Conference which is open to the general public. The conference is designed to focus attention of Forum members and the public on a limited number of issues significant to broad sectors of the industry. A Senior Management Forum is held annually attended exclusively by Forum members.

The Forum publishes a monthly magazine, *Nuclear Industry,* and distributes reports, monographs, and surveys. It also sponsors a Public Affairs and Information Program for the purpose of disseminating reliable information on all aspects of nuclear developments.

The Forum is governed by a 30-member Board of Directors and, during the intervals between meetings of the board, by a 9-member Executive Committee. The members of the board are selected from among the representatives of the Organization Members except for three members who may be scientists, educators, or other eminent persons who are not representatives of Organization Members.

The staff of the Forum is recruited from individuals with various professional backgrounds including business, engineering, economics, and law. The Forum's budget, which is financed primarily from dues paid by Organization Members and Associate Organization Members, has grown from about $100,000 in 1954 to a figure in excess of $600,000 in 1969.

The purposes, organization, and operation of the Forum

have been described in some detail because I feel that the forum concept, if applied in the case of the communications industry, might lead to the establishment of one or several organizations capable of playing significant integrating and coordinating roles in connection with the development of future public policies (1) by assembling relevant technical and economic information with regard to present and future types of services for home, business, and government, including alternative customer groupings with regard to such services; and (2) by studying and suggesting alternative solutions with regard to many of the technical, economic, and policy problems inherent in rational distributions of tasks and opportunities within the communications industry and between that industry and the information (data) processing industry, including particularly problems involving technical standards and compatibility of facilities.

While such information, studies, and recommendations will not in and of themselves resolve conflicts over distribution of tasks and opportunities, they will provide the kinds of rational input which may narrow the scope of such conflicts and thereby facilitate their resolution.

The quality of the information, studies, and recommendations will depend on the competence and integrity of the Forum staff and the leadership capabilities of individuals who will serve as chairmen of the Forum's committees, subcommittees, and working groups.

Whether or not one or several such organizations will be created will depend on (1) whether a few highly respected and prominently placed individuals in the communications and data processing industries and in government who are deeply concerned with future industry-government relations will exercise the necessary leadership; and (2) whether obvious constraints which stand in the way of creating such organizations can be overcome. One of the most obvious constraints is likely to be the scepticism on the part of the largest communications and computer companies as to whether the creation of such industry organizations will be in their best interests and the scepticism on the part of smaller companies whether the organizations would be dominated by the largest companies. Appropriate safeguards to allay both fears would have to be adopted in shaping such organizations. Whether such safeguards would be considered adequate will depend to a considerable extent on whether the disadvantages of *not* having available such organizations will appear to outweigh the risks entailed in creating them.

Whether such organizations, if they should be created, will be able to function adequately as interface devices between individual companies and industry segments in the private sector and regulatory agencies in the public sector will depend to a considerable extent on whether the organizations will be permitted to grow gradually in stature and competence or whether they will be overburdened at the beginning with details related to particular conflicts which happen to be "hot" at the time of their creation. The choice between sinking the organizations or permitting them to swim — slowly at first — will depend upon the members and the staff.

CHAPTER VI

Conclusions

At this point it is appropriate to remind the reader that the broad issue to which this study is addressed is the proper allocation of decision-making between the privately organized and the publicly organized segments of the communications system.

Rapid advances in communication and information technologies will make possible the furnishing of new communication and information services to the general public. On the other hand, other industries are going to step up greatly their demands for special bulk communication and information services and facilities. Statistics tend to indicate that the growth of special bulk business communication services and facilities has been far greater in recent years than the growth of services rendered to the general public.*

The present crossroads in the development of the communications system might be compared with that stage in the development of the transportation system when proprietary carriage by trucks began to increase sharply in relation to common carrier and contract carriage. A study of the economics of private truck transportation[1] concluded that proprietary trucking is a form of vertical integration on the part of firms which determine that the production of transportation services in addition to their primary products can make a positive contribution to total profits. Apart from the desire to exercise quality control and managerial control, this contribution to total profits is the principal reason for the

---

* See Appendix D.

growth of private carriage. The common carriers failed to check this growth because of their inability to compete with private trucking because of rate regulation. It will be interesting to see whether history will be permitted to repeat itself, and whether the regular communication carriers will be prevented from competing with private communications systems and specialized communication carriers.

It can be predicted without fear of contradiction that demands for bulk services and facilities, particularly on the part of companies which use computers for information gathering, storing, retrieving, processing, and distributing, will rise very sharply in the future. This circumstance plus the added circumstance that there will be an increasing overlap between technologies, facilities, and services concerned with transporting information and gathering, storing, retrieving, processing, and distributing information, will present great challenges for the structure-determining processes in two interdependent industries: the communications industry and the information processing industry.

Evidence of the increasingly uncertain borders separating the data processing industry from the communications industry may be found, for example, in the announcement of University Computing Company, in late October 1969, of the formation of a new wholly owned subsidiary, the Data Transmission Company. The company proposes to construct a $375 million microwave system designed to offer common carrier services in transmitting digital data exclusively. The company would be in direct competition with AT&T and Western Union. It is worth noting that during the preceding year University Computing Company was unsuccessful in its bid to take over control of Western Union.

Conflicts over rational distributions of task and opportunities within and between these industries and other industries which are large-scale users of communications and information services are bound to proliferate. Neither engineering nor economics nor any other single discipline is capable of furnishing the exclusive yardsticks to determine which resolution of these conflicts will assure that the communications system will meet in an acceptable manner the diverse needs which the system is expected to meet. A range of feasible solutions, however, can be established if the substantive criteria furnished by various disciplines are scrutinized with the aid of organizations and procedures reasonably well suited to such a task. The ultimate selection by a government agency of one of these solutions as being in the public interest then becomes a more manageable task than the task of developing singlehandedly and handing down the "right" decision without having been presented with a range of feasible alternatives. The recent proposed agreement worked out by the staffs of the National Association of Broadcasters and the National Cable Television Association, while perhaps deficient in some respects, constitutes a hopeful sign that some industry associations and some of their leaders stand ready to assume greater initiative in proposing resolutions of structural conflicts.

The assumption of such initiative is important because there is great danger of overloading the government on the intermediate political and the macro-political levels with demands for resolving structural conflicts. Such overloading can only be avoided, however, if companies in these industries and other industries which are large-scale users of communication services, individually and collectively, assume a fair share of this burden. This will require a greatly increased supply of personnel capable of acting not only as interpreters and integrators of different disciplines but also as interpreters and integrators of the privately organized and the publicly organized sectors of our political economy.

The growing importance of the roles of integrators both within companies which operate in areas of rapidly changing technologies, and between such companies and the government has been stressed by authors who have studied the business and political scenes.[2] Most of the knowledge and skill required for the discharge of those responsibilities will have to be acquired on the job. Greater freedom of movement between the two sectors will greatly aid the learning process. Academic institutions, however, can also assist materially in developing and transmitting some of the much needed general knowledge and skills required by such individuals. Academic compartmentalization, however, while effective in training specialists, has not been equally effective with respect to the development and transmission of knowledge and skills directed toward interpreting and integrating. Greater academic contributions could aid materially in allocating properly in the future decision-making functions between the private and public sectors of our political economy.

The failure of academic institutions to be effective interdisciplinary integrators has been stressed by Peter F. Drucker in his recent book, *The Age of Discontinuity:*

> . . . we will have to recognize that research produces information rather than knowledge, and to organize for application of information to end results, which is what we increasingly mean by knowledge, we need not one kind of person in the university, but many. Today's insistence on the Ph.D. for any job, that is, on the man with an advanced degree in a specialized discipline in which he has supposedly done research (that is, gathered information) is obscurantism. To be sure, we need people of this kind, but only a few of them in any area. The greatest need is for the man who can develop and teach the application to end results of knowledge and information drawn from diverse disciplines. We need further the man who can, in his own work, bring together knowledge and skills from a great many disciplines and integrate them into effective application outside the university. He is today not officially recognized — but he is the real "star" in today's large university.[3]

What, however, is likely to happen if company responses, individually and collectively, are not innovative but continue to follow traditional patterns? The first consequence is likely to be that the number of separate proceedings on the intermediate level, i.e., before the Federal Communications Commission, the courts, and specialized Congressional committees, will increase in number as individual companies attempt to bring about adjudications of individual conflicts in their favor. While company attorneys will continue to battle "enemies" in these proceedings, prolonged deadlocks will result preventing the companies concerned from proceeding with their respective long-range plans.

The danger of prolonged deadlocks is not limited to the communications systems. Other large politico-economic systems are already suffering from similar disabling conditions resulting from prolonged conflicts among multiple interest groups and government agencies. The electric power system, for example, experiences the consequences of prolonged deadlocks although the conditions which cause the conflicts are quite different. Here, too, the need for new mechanisms for speedier conflict-resolutions has been stressed.[4]

In the absence of such mechanisms which in the case of the communications system must be designed to bring about speedier resolutions of structural conflicts, top managements will continue gambling on the outcome of numerous battles and then hastily improvise courses of action designed to respond to whatever the outcome of such battles may be, instead of preparing alternative longer-range strategic plans based on determinations of company priorities to serve as negotiating positions in an attempt to reach settlements of structural conflicts.

The lines which will be drawn upon the conclusion of such battles are likely to be tactical rather than strategic. They are apt to prove quite resistant to change, however, as the contending parties entrench themselves behind those lines.

Instead of remaining flexible, therefore, and standing ready at all times to introduce new technologies and new services, companies are likely to become restricted to particular technologies and services.

Another consequence is likely to be that, given the absence of alternative plans submitted by industry, the macro-political level will be on its own in developing plans for distributing tasks and opportunities without the benefit of alternative, industry-developed plans. Companies therefore, individually and collectively, will find themselves in a position of reacting to governmental plans rather than assuming co-responsibility for their formulation. In other words, the decision-making functions will not be properly allocated between the private and the public sectors, and structural regulations are likely to be less flexible and workable than they need be. This, in turn, will affect adversely the long-range performance of the communications system, and will result in more vociferous demands for the government to take appropriate action.

The basis for these dire predictions is the course of events which has unfolded in the case of other large systems: transportation, health, and housing constitute a few representative examples. But this need not happen in communications. Fortunately, here time has not run out as yet.

APPENDIXES

REFERENCE FOOTNOTES

APPENDIX A

Questions Posed by the Federal Communications Commission with Respect to Future Cable Television Operations

(1) What is the appropriate relationship between CATV, communications common carriers, and other entities (e.g., the broadcasters, computer industry, etc.) which now provide, or may in the future seek to provide, communications services in the locality?

(2) What is likely to be the nature of the services that could be offered to the home or business under present and anticipated technology, and how would home and business requirements for communications facilities differ in light of services that might be economically practicable only for business use?

(3) Would the public interest be best served for the immediate future by:

(a) Permitting or encouraging the entry of all would-be newcomers, services, technologies, and facilities in an atmosphere of free competition, letting the market place determine the survival of the fittest, subject to such minimum regulation as may presently be required in the execution of the Commission's statutory responsibilities and to such future regulation as may become necessary or desirable in the public interest or as a result of legislation; or

(b) Permitting tests of different systems or services by different entities in various cities to afford some basis in experience for decisions as to the best ultimate structure before any particular system or service becomes established on a widespread basis; or

(c) Undertaking to devise a master plan now, before new facilities and services are inaugurated, to guide their development?

(4) Is it necessary or desirable that there should ultimately be

---

SOURCE: Notice of Inquiry and Notice of Proposed Rulemaking in Docket 18397, December 13, 1968, 15 F.C.C. 2d, p. 442.

a single cable (or bundle of cables) providing multiple means of communication to and from the home and/or business and, if so, should the complete system be owned by one entity or should there be diversity of ownership or control of some aspects of such a multipurpose communications system (e.g., joint ownership or indefeasible right of use)? What considerations should govern access to such system by communications common carriers and others offering communications services to the public? What should be the nature of the service offering by the entity or entities which would provide the cable (or bundle of cables) to the home?

(5) Is it necessary or desirable that there be multiple facilities providing means of communication to and from the home or business — e.g., some combination of radio, cable, and wire — and, if so, what kinds of services should in general be provided by what kinds of facilities?

(a) Is it technically and economically feasible for CATV to provide some two-way services, particularly two-way video, and switched services to and from the home and/or business and, if so, what would be the role of such services vis-à-vis other services such as videotelephone service?

(b) Assuming that some services could be provided by the facilities of more than one entity (by communications common carriers such as the telephone and telegraph companies, by CATV or some other enterprise), should duplication of facilities and competition in the provision of services be permitted, at least initially, or should there be some allocation of services among different entities?

(c) Assuming multiple facilities owned by different entities, would it be necessary or desirable to have a common junction at the premises of the consumer to facilitate interconnection of facilities and the provision of some services one way by one facility and the other way by another facility?

(d) Assuming multiple facilities owned or controlled by different entities, would it be necessary or desirable that the entire complex (or an essential portion thereof) be engineered according to uniform standards or by one entity to further technical compatibility, efficiency, and economy?

(6) What facilities would be necessary or desirable for trans-

mission through the streets, as opposed to from the street to the consumer's premises, and what are the comparative advantages or disadvantages of radio, cable, or some other mode?

(a) Should there be a variety of intracity distribution systems or only one, and if the latter, of what nature?

(b) Assuming a single intracity distribution system and a single cable (or bundle of cables) providing access to the premises of the consumer, should the complete system be owned by one entity or should there be diversity of ownership and control of some aspects? In either event, should there be limitations on common ownership or control of facilities in different cities?

(c) Apart from the question of ownership and control of facilities, should all entities desiring to provide a communications service to the public have nondiscriminatory and equitable access to the local distribution facilities for the purpose of so doing, and, if so, on what basis?

(7) How should the local communication system or systems tie into intercity terrestrial and satellite facilities?

(8) What technical standards would be necessary or desirable to achieve national and local compatibility and good quality service to the public?

(9) How could the same communications services available to homes in the city be provided to homes in rural or other areas not now economically reached by cable?

(a) To what extent could this problem be alleviated by the use of radio links such as those involved in the experimentation of Teleprompter Corp. and Chromalloy American Corp.?

(b) Would it be necessary or desirable for the Federal Government to subsidize construction of communications facilities in rural areas in a program akin to rural electrification?

(10) What should be the division of regulatory functions between Federal and State or local authorities with respect to the local communications system or systems: e.g., construction of facilities, terms and conditions of access by those offering communications services, services and charges to the public, licensing, etc.?

(a) Which aspects of the local system or systems would require uniformity and centralized regulation or would be important to the effectuation of national communications policies, which aspects would be primarily of local concern and appropriately subject to State or local regulation, and which aspects might better be left unregulated?

(b) What amendments to the Communications Act of 1934 might be necessary or desirable to effectuate the public interest and national communications policies in this area?

APPENDIX B

Additional Comments of the General Electric Company in the Matter of Establishment of Domestic Non-Common Carrier Communication Satellite Facilities by Non-Government Entities

General Electric's "Additional Comments" in the domestic satellite proceeding highlight this kind of problem. GE stated that the soaring costs of communication to meet GE's needs as a user have been a source of great concern to the company and have occasioned a thorough review within GE of this basic problem. GE proposed a major expansion and improvement in our national public communications system through the use of domestic satellites on a common carrier basis to render a number of new business record services: (1) "Telemail" (instantaneous business-to-business record communications), (2) remote access computer services, and (3) multi-access video services (for business meetings, coverage of seminars or meetings in academic or other institutions). GE suggested that separate specialized satellite systems might be established to meet the needs of the areonautical and broadcasting industries. GE stressed that the proposed system would concentrate on data and other record message transmissions which would not require the availability of complex switching arrangements. System costs required fully to implement such a system by 1980 were estimated to be $321 million, of which $140 million would be the cost of earth facilities.

After detailing these plans, GE stated that it had concluded that prudent business judgment would not warrant the company's taking

---

SOURCE: FCC, Docket No. 16495, February 19, 1969.

. . . affirmative action that would directly, on an investment and operational basis, involve it in an undertaking to bring about the advantages of the satellite services enumerated.

In short, it is impossible to judge with even reasonable certainty the likely nature of the entity ultimately lodged with authority to establish a domestic satellite system or the extent to which, if at all, participation by a company such as General Electric would be possible. Additional critical factors, of course, relate to the uncertainty of the time that will elapse before such matters are known and the resulting inability to assess the probable nature of GE's total business commitment at that time. It is, however, these very fundamental matters which must be taken into account in any necessary business judgments. Lacking a further delineation of policy the imponderables remain speculative at best.

The General Electric Company has, therefore, found that it could not in the exercise of a prudent business judgment propose to undertake commitments of an investment or operational nature. It continues to stand ready, however, to provide technical and other data that may assist the Commission in arriving at the required public policy determinations, and it is in this context that this document has been submitted.

Such business determinations must proceed from an evaluation of total existing and foreseeable commitments taking into account a multiplicity of factors specific to the company involved and the venture contemplated. In approaching the domestic satellite field it was concluded that there is no realistic way in which to determine, from the standpoint of a potential business commitment, the manner in which the field may develop. For instance, it is entirely possible that the eventual domestic satellite ownership pattern may be so structured as to foreclose or discourage direct investment from private enterprise. The uncertainty of the structuring of a future domestic common carrier satellite system is further clouded when it is recognized that the final product will likely be subject to the actions of not only the Commission, but of Congress, the Executive, and other appropriate authorities.

APPENDIX C

Apportioning the Communications Systems

In distributing tasks and opportunities within the communications system, the conceptual framework of "apportioning" and "re-apportioning" that system seems appropriate because:

1. It accommodates reasoned arguments advanced by various disciplines for or against the drawing of particular lines without arousing strong pro or anti sentiments;
2. It focuses broadly on the present and future communications system as a whole rather than parts of the system;
3. It promotes the balancing rather than the polarizing of those arguments, thereby promoting dialogues which are likely to generate trade-offs and agreements;
4. It emphasizes the notion that distributions of tasks and opportunities are subject to modification as technologies and other conditions change;
5. It has political overtones which highlight the fact that structure-determining processes are political processes to a considerable degree;
6. It distinguishes "apportioning" from other types of regulations and thus tends to promote more ready acceptance of informal procedures in connection with "apportioning" processes.

APPENDIX D

GROWTH OF REVENUES OF TELEPHONE COMPANIES BY SELECTED CLASSES OF SERVICE

(Figures Expressed as an Index Using 1946 as a Base)

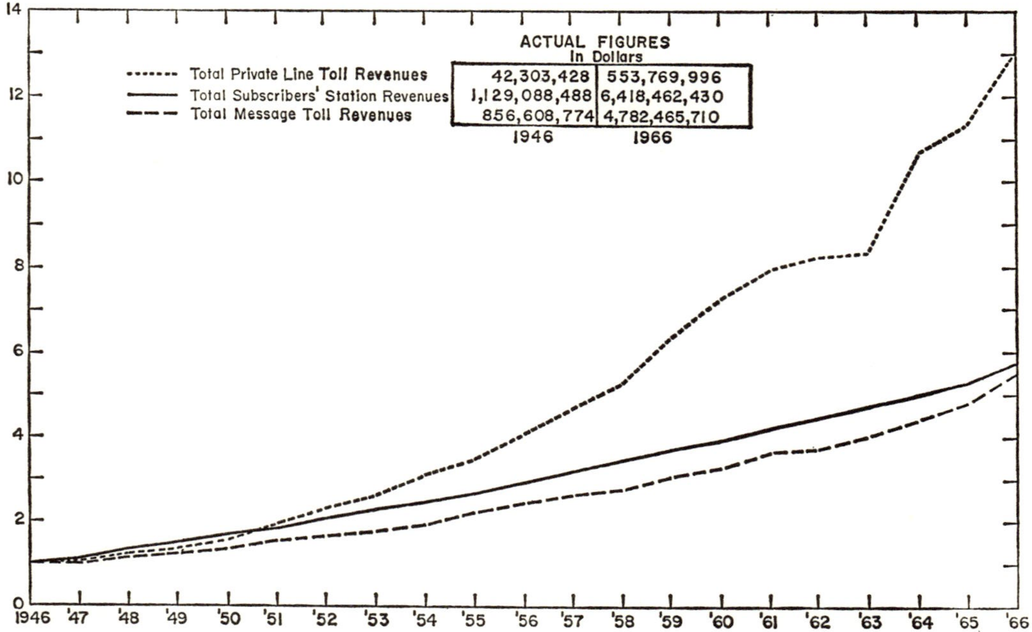

Actual Figures Taken from FCC Common Carrier Statistics

REFERENCE FOOTNOTES

Chapter I

[1] Emmette S. Redford, *American Government and the Economy* (New York, The Macmillan Company, 1965), p. 32.

[2] U.S. Department of Commerce, National Bureau of Standards, Institute for Applied Technology, February 1969, PB 183612.

[3] Given at the Regional Conference on Issues and Priorities for the Development of the American Northwest, Seattle, June 23, 1969.

[4] See also Carl Kaysen, "Model-Makers and Decision-Makers: Economists and the Policy Process," *The Public Interest,* No. 12, Summer 1968, p. 80.

[5] Data for calendar year 1966 taken from Federal Communications Commission (FCC), *33rd Annual Report for the Fiscal Year 1967.*

[6] See particularly Marver H. Bernstein, *Regulating Business by Independent Commission* (Princeton, Princeton University Press, 1955); Louis M. Kohlmeier, Jr., *The Regulators, Watchdog Agencies, and the Public Interest* (New York, Harper and Row, 1969); and Emmette S. Redford, *op. cit.*

[7] The term "polycentric" was suggested by Michael Polanyi in *The Logic of Liberty* (Chicago, University of Chicago Press, 1951). In his final chapter entitled "Manageability of Social Tasks" (pp. 154 ff.) Polanyi argues that polycentric conflicts can be socially managed only through mutual adjustments and not by central direction.

[8] See Concurring Statement of Chairman E. William Henry, Docket No. 16258, December 23, 1965.

[9] See Douglas Needham, *Economic Analysis and Industrial Structure* (New York, Holt, Rinehart and Winston, 1969); Joe S. Bain, *Industrial Organization* (New York, John Wiley & Sons, 2d ed., 1968); Richard Caves, *American Industry: Structure, Conduct, Performance* (Englewood Cliffs, Prentice-Hall, 1964).

Chapter II

[1] Federal Communications Commission, "Special Investigation Docket No. 1," Exhibit 2096-A (June 13, 1937), Appendix 7, Sheet No. 31.

[2] *Report of the Telephone and Telegraph Committees of the Federal Communications Commission in the Domestic Telegraph Investigation,* Docket No. 14650, April 29, 1966, p. 190.

[3] Harry B. MacMeal, *The Story of Independent Telephone* (Independent Pioneer Telephone Association, 1934).

[4] FCC, *34th Annual Report for the Fiscal Year 1968.*

[5] Willis-Graham Act of June 10, 1921 (42 Stat. 27).

[6] *U.S.* v. *GT&E,* Civ. 1912 (S.D.N.Y., June 19, 1964).

[7] FCC, Docket Nos. 16942 and 17073.

[8] FCC, "Memorandum Opinion and Order," Docket Nos. 16942 and 17073, adopted June 26, 1968 (FCC 68-661).

[9] FCC, *32nd Annual Report for the Fiscal Year 1966,* p. 44.

[10] Report and Order adopted on July 29, 1959, affirmed upon reconsideration in a Report and Order adopted on September 28, 1960.

[11] Figures in this section were furnished by the FCC Safety and Special Radio Services Bureau.

[12] Quoted in *Telecommunications Reports,* Vol. 35, No. 29, July 14, 1969, p. 24.

[13] FCC, *34th Annual Report for the Fiscal Year 1968,* pp. 146–147.

[14] FCC, *32nd Annual Report for the Fiscal Year 1966,* pp. 48–52.

[15] FCC, Docket No. 8658.

[16] *Telecommunications Reports,* Vol. 34, No. 36, August 26, 1968.

[17] Communications Act of 1934, Sec. 222 (c) (2).

[18] Communications Act of 1934, Sec. 222 (b) (1).

[19] Memorandum Order and Opinion of March 17, 1964 (FCC 64-217), 37 FCC Report, p. 1151.

[20] Communications Satellite Act of 1962, Sec. 102 (a) and (c).

[21] For a summary of the 30 circuit situation, see House Report 1836, 90th Congress, 2nd Session, August 1, 1968, "Government Use of Satellite Communications, 1968," 34th Report by the Committee on Government Operations, U.S. House of Representatives.

[22] FCC 68-212, February 18, 1968.

Chapter III

[1] *The First 25 Years of RCA: A Quarter of a Century of Radio Progress* (New York, Radio Corporation of America, 1944). Also Eugene Lyons, *David Sarnoff, A Bibliography* (New York, Harper and Row, 1966).

[2] Report of the Federal Trade Commission on the Radio Industry in response to House Resolution 548, 67th Congress, 4th Session, 1923, p. 83.

[3] *U.S.* v. *Southwestern Cable Co.,* 390 U.S. 157 (1968).

[4] The Supreme Court so construed present copyright law in *Fortnightly Corporation* v. *United Artists,* 392 U.S. 390 (1968). Congress has under consideration a general revision of the copyright law, including provisions relating to copyright liabilities of CATV systems.

[5] Notice of Proposed Rulemaking and Notice of Inquiry, Docket No. 18397, December 12, 1968 (FCC 68-1176), 15 FCC 2d 417.

[6] Ibid., p. 69.

[7] The FCC has listed the following predicted services: ". . . facsimile reproduction of newspapers, magazines, documents, etc.; electronic mail delivery; merchandising; business concern links to branch offices, primary customers or suppliers; access to computers; e.g., man-to-computer communications in the nature of inquiry and response (credit checks, airlines reservations, branch banking, etc.), information retrieval (library and other

reference material, etc.), and computer-to-computer communications; the furtherance of various governmental programs on a Federal, State, and municipal level; e.g., employment services and manpower utilization, special communications systems to reach particular neighborhoods or ethnic groups within a community, and for municipal surveillance of public areas for protection against crime, fire detection, control of air pollution, and traffic; various educational and training programs; e.g., job and literacy training, preschool programs in the nature of "Project Headstart," and to enable professional groups such as doctors to keep abreast of developments in their fields; and the provision of a low-cost outlet for political candidates, advertisers, amateur expression (e.g., community or university drama groups), and for other moderately funded organizations or persons desiring access to the community or a particular segment of the community." The Commission stated that the list was not intended to be all-inclusive. Ibid., p. 420.

[8] See p. 31 above.

[9] *Telecommunications Reports,* Vol. 35, No. 31, July 28, 1969, p. 9.

[10] *Wall Street Journal,* August 27, 1969.

[11] *Connecticut Committee Against Pay-TV* v. *FCC,* 301 F. 2d, 835 (C.A.D.C., 1958) cert. den., 371 U.S. 816.

[12] Fourth Report and Order, Adopted December 12, 1968, Docket No. 11279, 15 FCC 2d 466.

[13] *Weaver* v. *Jordan,* 49 Cal. Rptr. 537, 411 P (2d) 289 (1966), cert. den. 385 U.S. 844.

Chapter IV

[1] The basis for this approach is to be found in Harold Laswell, *Politics — Who Gets What, When, How* (New York, McGraw-Hill Book Company, 1936; republished by the World Publishing Company, New York, 1958).

[2] Paul R. Lawrence and Jay W. Lorsch, *Organization and Environment: Managing Differentiation and Integration* (Boston, Division of Research, Harvard Business School, 1967).

[3] U.S. Department of Labor Bulletin No. 1596, *1967 Directory of National and International Labor Unions in the United States.*

[4] The waves are measured in terms of their length (measurements are expressed in meters) and their frequency (measurements are expressed in cycles per second). Wave-length and frequency stand in an inverse relationship to each other because if multiplied the product is constant, being the speed of light (300 million meters per second).

[5] Communications Act of 1934, Sec. 301. "It is the purpose of this Act, among other things, to maintain control of the United States over all channels of interstate and foreign radio transmission; and to provide for the use of such channels, but not the ownership thereof, by persons for limited periods of time, under licenses granted by Federal authority, and no such license shall be construed to create any right, beyond the terms, conditions, and periods of the license."

[6] FCC, *32nd Annual Report for the Fiscal Year 1966,* p. 50.

[7] At present, regulatory authority with regard to uses of the radio spectrum is divided between the President with respect to Federal government uses and the Federal Communications Commission with respect to other uses (private uses and uses by states and local governments). The President has delegated his authority to the Director of Telecommunications Management. The division of spectrum uses between Federal government and other uses is determined by informal agreements between the Director and the FCC. In allocating frequencies, the Commission and the Director abide by frequency allocations made internationally through agreements concluded under the auspices of the International Telecommunication Union, an organ of the United Nations. The United States and over 120 other nations are members of the International Union.

[8] See William H. Meckling, "Management of the Frequency Spectrum," *Washington University Law Quarterly,* Vol. 1968, No. 1, Winter 1968, p. 26; and Harvey J. Levin, "The Radio Spectrum: Economic-Physical Character and Regulatory Framework," *The Journal of Law and Economics,* Vol. XI, No. 2, October 1968, pp. 433–501. Both papers were commissioned by Resources for the Future, Inc. and the Brookings Institution.

[9] Stanley D. Metzger and Bernie R. Burrus, "Radio Frequency Allocation in the Public Interest: Federal Government and Civilian Use," *Duquesne University Law Review,* Vol. 4, No. 1, 1965–66, pp. 1–96. The authors state in a footnote that their study was conducted under contract with Motorola, Inc., one of the principal manufacturers of mobile communications equipment, but that they enjoyed complete freedom to conduct this study as they saw fit, and that, in consequence, Motorola is not responsible for the analyses and conclusions contained in the study.

[10] The Bureau of the Budget, at the request of President Johnson, has undertaken a study of Federal agencies which discharge various regulatory responsibilities regarding communications.

[11] The existence of such a communications gap was demonstrated at a conference on the use and regulation of the radio spectrum sponsored jointly by the Brookings Institution and Resources for the Future, Inc. and held on September 11 and 12, 1967, at Airlie House, Warrenton, Virginia. The author was permitted to attend the conference as an observer. For comments on that conference and the communications gap, see William K. Jones, "Use and Regulation of the Radio Spectrum: Report on a Conference," *Washington University Law Quarterly,* Vol. 1968, No. 1, Winter 1968, p. 71.

[12] For a good discussion of such efforts in the case of the communications industry, see Donald F. Turner, "Role of Antitrust Policy in Communications Industry," *The Antitrust Bulletin,* Vol. XIII, Fall 1968, p. 873 ff.

[13] Robert A. Levine, "Rethinking our Social Strategies," *The Public Interest,* No. 10, Winter 1968, p. 86.

[14] For a discussion of this phenomenon, see Raymond A. Bauer, Editor, *Social Indicators* (Cambridge, The MIT Press, 1966), p. 35.

[15] *FCC* v. *RCA Communications, Inc.,* 346 U.S. 86, 97L, ed. 140, 73 S. Ct. 998 (1953).

[16] *Microwave Communications, Inc.,* 18 FCC 2d 953, at p. 971.

[17] In this connection it is interesting to note that Pan American Airways has instituted subterminals in New York City suburbs from where passengers are transported by small buses directly to the planes at Kennedy Airport. This constitutes a recognition on the part of Pan Am that the public seeks a closer approximation to door-to-door long-distance transportation services than are presently provided by separate companies engaged in providing various elements of such services. Similarly, American Airlines and American President Lines have formed an "air bridge" which will permit a shipper to move products in a new container that would be compatible with both air and sea movement. A single document at a joint freight rate would facilitate door-to-door movements of such containers.

[18] See a sequence of articles by Robert H. Bork and Ward S. Bowman, Jr., taking the former position and Harlan M. Blake and William K. Jones, taking the latter position. See "The Goals of Antitrust — A Dialogue on Policy," *Columbia Law Review,* Vol. 65, 1965, p. 363 ff.

[19] Sec. 408 (a) of the Civil Aeronautics Act. Vertical integration between airline operations and aircraft manufacturing was first outlawed by the Air Mail Act of 1934. That act prevented persons holding air mail contracts from having any interests in companies engaged in other phases of aeronautics. That prohibition was carried over into the Civil Aeronautics Act which makes unlawful without approval by the Civil Aeronautics Board mutual control of any air carrier and any firm "engaged in any phase of aeronautics otherwise than as an air carrier." The Board may approve such arrangements if they are found to be in the public interest, provided they do not result in a monopoly or monopolies or jeopardize another air carrier not party to the agreement (Sec. 408 (b)).

[20] Richard E. Caves, *Air Transport and Its Regulators* (Cambridge, Harvard University Press, 1962).

[21] *Wall Street Journal,* May 12, 1969, quoting from CAB decision: "The basic solution to the industry's present financial situation would appear to lie in exercising restraint in ordering new flight equipment and in the use of its available capacity, rather than in increasing its price to the public."

[22] This was pointed out with regard to the nuclear power supply industry in a study prepared by Arthur D. Little, Inc., *Competition in the Nuclear Power Supply Industry,* Report to the U.S. Atomic Energy Commission and the U.S. Department of Justice, November 1, 1968, Part I, p. 22.

[23] Robert Schlaifer and S. D. Heron, *Development of Aircraft Engines; Development of Aviation Fuels; Two Studies of Relations Between Government and Business* (Boston, Division of Research, Harvard Business School, 1950), p. 14.

[24] *Telecommunications Reports,* Vol. 35, No. 28, July 7, 1969, p. 1.

[25] For a general discussion of the role of adjudication in conflict resolution

see Harold J. Berman and William R. Greiner, *The Nature and Functions of Law* (Brooklyn, The Foundation Press, Inc., 2d ed., 1966), and Lon L. Fuller, *The Forms and Limits of Adjudication,* reproduced but unpublished manuscript, Harvard Law School Library, 1959.

[26] In the Matter of American Telephone and Telegraph Company, "Foreign Attachment," Tariff Revisions, AT&T Tariff FCC Nos. 263, 260, and 259, Memorandum Opinion and Order, Adopted December 24, 1968 (FCC 68-1234).

[27] *Telecommunications Reports,* Vol. 35, No. 28, July 7, 1969.

[28] The National Academy of Sciences, *A History of First Half-Century of the National Academy of Sciences, 1863–1913,* Washington, 1913.

[29] *Telecommunications Reports,* Vol. 35, No. 31, July 28, 1969, p. 8.

[30] For a critique of such expectations see particularly Marver H. Bernstein, *Regulating Business by Independent Commission* (Princeton, Princeton University Press, 1955), and Louis M. Kohlmeier, Jr., *The Regulators, Watchdog Agencies and the Public Interest* (New York, Harper and Row, 1969).

[31] The need for openness of informal processes as a check upon arbitrariness has been stressed by Kenneth Culp Davis, *Discretionary Justice — A Preliminary Inquiry* (Baton Rouge, Louisiana State University Press, 1969), pp. 111 ff.

[32] Letter from Richard W. McLaren, Assistant Attorney General, Antitrust Division, to Malcolm W. Jensen, Chief, Division of Weights and Standards, National Bureau of Standards, Department of Commerce, referred to in *The New York Times,* March 18, 1969.

Chapter V

[1] For a general discussion of this subject see Emmette S. Redford, *American Government and the Economy* (New York, The Macmillan Company, 1965.)

[2] *Telecommunications Reports,* Vol. 35, No. 34, August 18, 1969, p. 16.

[3] *Telecommunications, A Program for Progress,* A Report of the President's Communications Policy Board, March 1951. The Board, which was established in February 1950 by President Truman (Executive Order 10110, February 17, 1950), consisted of Dr. Irwin Stewart, Chairman, Lee A. DuBridge, William L. Everitt, James R. Killian, Jr., and David H. O'Brien. It is worthy of note that Dr. DuBridge is at present President Nixon's Science Advisor and that Dr. DuBridge has been directed by the President to review a Communications Task Force report which was submitted to President Johnson in 1968.

[4] Report of a Special Advisory Committee on Telecommunications, December 20, 1958. The Committee consisted of Victor E. Cooley, W. Preston Corderman, Frank G. Kear, Irwin Stewart, and William G. Thomas and was appointed in the Office of Civil and Defense Mobilization on November 20, 1958.

[5] Executive Order 10995, February 19, 1962 (27FR 1519).

[6] Raymond A. Bauer, in a chapter entitled "Application of Behavioral Science," contained in a Report to the Committee on Science and Astronautics, U.S. House of Representatives, by the National Academy of Sciences, June 1967, contends that negotiation, whether explicit or implicit, constitutes the basic process by which the heterogeneity of values and interests in a complex society is solved. Charles E. Lindblom, in *The Intelligence of Democracy* (New York, The Free Press, 1965) goes one step further in claiming that decision making through mutual adjustment as a decision-making process is superior to other decision-making processes, and particularly to the planning, programming, and budgeting system which aims at reaching optimal decisions.

[7] Douglas Needham, *Economic Analysis and Industrial Structure* (New York, Holt, Rinehart and Winston, Inc., 1969), p. 137.

[8] For an extended discussion of "continua" of means to achieve public objectives, see Robert A. Dahl and Charles E. Lindblom, *Politics, Economics, and Welfare* (New York, Harper and Row, 1963; originally published in 1953 by Harper and Brothers).

[9] Tom Alexander, "The Unexpected Payoff of Project Apollo," *Fortune,* July 1969, p. 114. See also James E. Webb, *Space Age Management — The Large-Scale Approach* (New York, McGraw-Hill Book Company, 1969).

CHAPTER VI

[1] Walter V. Oi and Arthur P. Hurter, Jr. *Economics of Private Truck Transportation* (Dubuque, Iowa, Wm. C. Brown Co., 1965).

[2] Paul R. Lawrence and Jay W. Lorsch, *Organization and Environment — Managing Differentiation and Integration* (Boston, Division of Research, Harvard Business School, 1967), have pointed out this need on the intra-company level. Charles E. Lindblom, in *The Policy-Making Process* (Englewood Cliffs, Prentice-Hall, 1968) has made similar observations with regard to the intercompany and government-industry levels: "A critically important policy-making specialist . . . is the broker, the arranger, the peacemaker. . . . [Parties] left to themselves, would not always have recognized their common interests. In an extreme, a broker is not much interested in whether this or that policy wins: he is concerned only that some policy emerges, that the political work of policy making gets done." (p. 99)

[3] Peter F. Drucker, *The Age of Discontinuity — Guidelines to Our Changing Society* (New York, Harper & Row, 1969), p. 355.

[4] *The New York Times,* July 28, 1969, reported that the outgoing chairman of the Federal Power Commission, Lee C. White, stressed the need for mechanisms for speedy resolution of disputes over the location and design of power plants and power lines. He pointed out that these disputes on local and state levels which generally pit conservation interests against the power industry result in long delays in the construction of needed power facilities leading to threatened "brown-outs" during periods of peak demand. Mr. White proposed giving the Federal government more authority to re-

solve such disputes, but he felt that the industry was so fearful of the government that it fought the proposed legislation successfully. He stated that his greatest regret was that the industry did not even come up with a counterproposal of its own. He predicted that in view of the critical nature of potential shortages of power generation and transmission facilities, the industry "will be in Washington within two years demanding legislation to help them meet their responsibilities for supplying adequate and reliable electric service throughout the country."